EATING MAY BE
HAZARDOUS
TO YOUR HEALTH

Other books by Jean Carper

STAY ALIVE!

BITTER GREETINGS

THE DARK SIDE OF THE MARKETPLACE
  (co-author with Senator Warren G. Magnuson)

NOT WITH A GUN

Eating May Be Hazardous to Your Health

Jacqueline Verrett and Jean Carper

Anchor Books
ANCHOR PRESS/DOUBLEDAY
GARDEN CITY, NEW YORK
1975

EATING MAY BE HAZARDOUS TO YOUR HEALTH
was originally published in hardcover by Simon & Schuster, Inc.,
in 1974.

Anchor Books Edition: 1975
ISBN: 0-385-11193-2

In memory of
CHUNIE

CONTENTS

INTRODUCTION

Bad and Getting Worse

"The thing that bugs me is that the people think the FDA is protecting them. It isn't. What the FDA is doing and what the public thinks it's doing are as different as night and day."

That statement would not be so remarkable had it not been made by Dr. Herbert Ley, Jr., a short time after he was relieved of his job as the Commissioner of the Food and Drug Administration. According to his superiors, Dr. Ley had goofed in handling the cyclamate (a noncaloric sweetener) situation, and he was ousted in December 1969. But since that time, although the FDA has been confronted with one crisis after another involving food additives, the furor has died down; personnel stay despite their mistakes. A cyclamate-like crisis that once shook the whole Department of Health, Education and Welfare, and most of the country, now hardly causes a tremor of concern.

If anything, the situation today is worse, precisely because we now know much more about the dangers of food additives, and less protective action is being taken. It seems that our government protectors have simply dug in, hoping to weather out what an FDA official once called the "dangers of rampant consumerism."

In actual fact, these "dangers of rampant consumerism" regarding food additives come from but a few informed individuals and groups who constantly do battle with federal agencies and officials who they believe are not sufficiently protecting the public from food pollution. The 210 million Americans, on the other hand, who have an enormous stake in the outcome, are hardly aware of the real issues or the risks they encounter as a result of government policy on food chemicals. Some persons in fact believe any hazard from additives in food to be remote and perhaps faintly ridiculous. How could such tiny doses of chemicals in food really harm you? And how could the government let you eat them if they were dangerous? Though the presumption that your government is protecting you is natural enough, it's unfounded, as Dr. Ley pointed out, and as you'll see illustrated throughout this book.

Nor is concern over food additives an alarmist point of view held by a few organic-food faddists who are against anything technology has to offer. Though such crusaders have popularized a current movement against hidden chemicals in foods, they can't be dismissed as a lunatic fringe. Even industry-oriented *Fortune* magazine, in an article titled "Hysteria about Food Additives," grudgingly conceded: "The anti-additive camp clearly has its share of cranks, conspiracy theorists, and exaggerators for effect, but there is also a deep and serious logic behind their crusade. Plenty of thoughtful scientists are concerned about the proliferation of strange new chemicals to which humans are exposed these days, and not only in food."[1] True. Some of the country's most respected and prestigious scientists are deeply disturbed about the poisonous aspects of chemicals in our food—scientists such as Nobel-laureate geneticist Joshua Lederberg; Dr. Wilhelm Hueper, former chief of the National Cancer Institute's environmental-cancer section; Dr. Umberto Saffiotti, associate director for carcinogenesis at the National Cancer Institute; and Dr. Samuel Epstein, cancer researcher and professor at Case Western Reserve University. Senators Edward M. Kennedy, Gaylord Nelson and Abraham A. Ribicoff, as well as Representative H. L. Fountain, have sponsored legislation and held numerous hearings to try to clean up our food supply. Ralph Nader has a number of staff members, including the Health Research Group, who dog the FDA and the U. S. Department of Agriculture to end what they consider the illegal uses of dangerous chemicals in food. And there are a number of unheard-from scientists within these agencies who are alarmed about the reckless use of such chemicals.

This book is written by two sane, we hope, and sober individuals, who are not "food nuts" but are concerned about the health of their bodies and those of other Americans who are unknowingly consuming a multitude of food chemicals. Dr. Jacqueline Verrett is an insider—a biochemist and a researcher with the FDA for fifteen years. Jean Carper is a consumer-science writer who has followed the activities of the FDA since the cyclamate incident. Whenever the first-person "I" is used in the book, it refers to Dr. Verrett, who must be considered the principal authority because of her scientific expertise and firsthand association with the FDA.

To those "dangerous advocates of rampant consumerism," some of whom should be mentioned by name—Ralph Nader; James S. Turner, author of *The Chemical Feast* and director of Consumer Action for Improved Foods and Drugs; Anita Johnson and Dr. Sidney Wolfe, Health Research Group; Harrison Wellford, Center for Study of Responsive Law; David Hawkins, National Resources Defense Council, Inc.; Peter Schuck, Consumers' Union, Washington office; Ruth Desmond, Federation of Homemakers, Inc.; Michael F. Jacobson, Ph.D., Center for Science in the Public Interest; Dr. Epstein —there may be little in this book that they do not know already or have not themselves contributed to making public.

But it is our hope that this book will lead the consumer who has not had the opportunity for such awareness, through the labyrinth of government involvement in food chemicals and leave him aware and convinced that he should be not only concerned but *alarmed* about the chemicalization of his food and the government's failure to do something about it.

JACQUELINE VERRETT
JEAN CARPER

I

What's Wrong with Your Food?

1

What Your Government Doesn't Tell You

First, the good news: The federal government, mainly the Food and Drug Administration of the Department of Health, Education and Welfare and to a lesser extent the U. S. Department of Agriculture, is empowered to keep your food safe for consumption and free of dangerous chemicals. Now the bad news: They do nothing of the kind. As a result our food supply is permeated with chemicals of dubious safety.

It's not simply a matter of

> water, syrup, shortening, sugar, whey solids, food starch modified, dextrose, sodium caseinate, flavoring gelatin, whole milk solids, monosodium and diglycerides, salt, vinegar, polysorbate No. 60, vanilla, monosodium phosphate, guar gum, lecithin, artificial color in a crust of wheat flour, sugar, shortening, water, dextrose, graham flour, sorghum grain flour, salt, sodium bicarbonate, ammonium bicarbonate, artificial flavoring and coloring—

better known as lemon cream pie; or of

> water, sugar, nonfat dry milk and whey solids with calcium hydroxide and disodium phosphate, modified tapioca starch, hydrogenated vegetable oil, cocoa processed with alkali, emulsifiers (sodium stearoyl-2-lactylate, polysorbate 60, and sorbitan monostearate), artificial color and flavor, sodium caseinate, salt, dextrose, carrageenan, guar gum—

chocolate pudding.

Such chemical foods are now quite familiar to you if you just glance at labels. You may have even become inured to the lists that run on like endless entries in a chemical dictionary. What can it matter? The food may taste good. It's convenient. And surely, you reason, your government would not permit you and 210 million others to eat it if it weren't safe.

But also—what about peanut butter and hot dogs and milk and bacon and potatoes and corn and eggs and beef? These are not mere food analogues concocted in a laboratory. These are real foods which you should be able to depend on not to be harmful. These are staples, basic foods, life-sustaining nutrients that you and your family eat every day.

These foods, too, are often not safe. Some knowledgeable scientists have given up eating hot dogs, bacon, ham, sausages, anything containing the chemical nitrite, after seeing how this chemical, interacting with others, has left laboratory animals grotesquely deformed with cancers. Peanuts, as well as peanut butter, corn, and other grains that go into bread and beer sometimes are contaminated with aflatoxin (a natural chemical produced by certain molds), one of the most potent cancer-causing agents ever discovered. Eggs and chickens may contain PCBs, industrial chemicals that have caused poisonings in Japan. Many foods, especially carbonated noncola beverages and gelatins, are infused with a red dye that in animals has caused birth defects and killed unborn fetuses. Monosodium glutamate (MSG) is still widely used in all kinds of processed foods, though it has been shown to cause brain damage in infant animals. Blighted potatoes have been implicated in birth defects such as spinal and head malformations. And ad infinitum—or, more appropriately, ad nauseam.

That things are this bad is not something your government is officially likely to tell you. There is not likely to be a press release from the FDA's public information office tomorrow saying: "Nothing's left that's fit to eat," or "Whole population being slowly poisoned." Nor is an official likely to suggest slapping a warning label on foods similar to that familiar to cigarette smokers: "Eating May Be Hazardous to Your Health." After all, in most cases it was the government that allowed things to get into such a sorry state, and it does have a self-protective stake in the affair.

Yet, imagine how much easier it would be if the government would just out with the truth. It would end all that intrigue, all the time-consuming hard work trying to cover up mistakes, juggling scientific data to make it come out right, always assuming a defensive posture, evading responsibility, playing footsie with the industry, assuring everyone that no matter what a few die-hard scientists say and do, there is no need for concern, that our food is the safest in the world and there's not a shred of evidence that anyone has ever been harmed by eating an additive.

Unfortunately, our food is not the safest in the world; at least, some other countries think it isn't. Sweden, for example, refused to import U.S. beef that had been raised on feed containing diethylstilbestrol (DES), a growth-promoting hormone known to cause cancer. Twenty-one nations, in fact, banned DES in livestock before this country got around to banning it in 1973 after much delay. Norway has outlawed the use of nitrites in certain food. Great Britain prohibited brominated vegetable oils in 1970 (they're still allowed here). In general the United States has a very relaxed attitude toward chemicals in foods compared with other countries, which follow a much more conservative policy. The Soviet Union is especially strict, allowing only a minimum of chemicals in food and quickly banning those found to be potentially harmful.

Notwithstanding the FDA's proclamations to the contrary, all is not right with our food supply and we had best do something about it. What your government is not telling you can hurt you, and unless there is public pressure the FDA and the USDA will not be swayed from a course they are now taking that in the future will expose us to even more potential danger —without our knowledge or consent.

Why You Can't Protect Yourself

If you stop to think about it, we have more free choice about drugs than about the food chemicals we eat. And remember, many food chemicals are not far afield from drugs. People who wouldn't think of medicating themselves daily with a hundred or more different drugs take in that many food chemicals whose biological action on cells may be as potent as that from drugs. Some food additives, in fact, double as drugs —for example, both nitrite and MSG were used as drugs—and animal feed is filled with innumerable antibiotics which end up as residues in meat, milk and eggs we eat. The line is thin: Thalidomide was a drug, cyclamate was a food additive, yet biologically their action on certain living tissue was nearly identical. Both produced the dreaded "flippers" in baby chicks. Thalidomide, we know, also did the same in humans; we have no sure way of knowing about cyclamate.

We complain constantly that we have become a drug culture—with kids dosing themselves with LSD, marijuana and heroin, adults wolfing down barbiturates, tranquilizers, amphetamines and hundreds of other supposedly therapeutic concoctions. But few people also point out that we have become a nation of "food junkies."

Even dying persons have a choice about whether they want to take an experimental drug that may kill or cure them. A doctor explains the possible risks and benefits, and the patient can take it or leave it. In medical terminology, the decision to do so is called "informed consent." But we as Americans, dependent on a vast network of mass-produced, processed food, cannot take it or leave it. We don't have the rights of "informed consent" even about chemicals which may be of no value whatsoever to us. The Food and Drug Administration, as the Nation's Physician—the biggest doctor of them all— makes that decision for us every day. And every day at the FDA's discretion we put into our mouths hundreds of chemicals that are biologically potent but about which we know virtually nothing. As Senator Gaylord Nelson has said, "We are being chemically medicated against our will."

If some food additives were regulated as drugs they would be forbidden—except by prescription, and then forced to carry warnings—especially to pregnant women. (As a result of the thalidomide tragedy, all prescription drugs routinely carry warnings to physicians about consumption during pregnancy.) Some food additives might even be considered too dangerous as prescription drugs, which, even though therapeutic, are not allowed unless their benefits outweigh the risk; drugs that cause cancer, for example, can't be dispensed willy-nilly. And surely over-the-counter drugs available to everyone (which is essentially what food additives are) can't be approved if they carry dire threats to our health.

Such is the FDA's double standard that with one hand it regulates drugs while with the other it allows a whole nation of women to eat as much Red 2, a food coloring hidden in many food products, as it wants. There is no official recognition or warning that this chemical caused stillbirths, fetal deaths and birth defects in animals. All of us presumably healthy people, who don't need or want such chemicals for any specific health purpose, nevertheless ingest them, though many have never been tested and have no therapeutic value and in some cases no value whatsoever to us, even economic.

What chance do you have of detecting and rejecting dangerous additives on your own? Very little. When some persons hear of dangers in a certain food additive, they check food labels, and if that ingredient is listed they give up eating that food. You might get some limited protection that way. Some persons who heeded advance warnings about cyclamates stopped drinking diet soft drinks and were undoubtedly better off for it. But do-it-yourself protection is folly—inefficient and often impossible. You might as well try wearing a metal shield to protect yourself against radiation from a new atomic reactor plant set up in your back yard. Or an oxygen mask against air pollution. Food additives, like radiation and environmental hazards in general, present problems that must be solved by governmental action.

You can't even always tell which chemicals have been deliberately added to foods, let alone those that have slipped in accidentally, because labeling regulations are inadequate. A manufacturer does not have to list constituent chemicals on a label if that chemical is either "mandatory" or "permissible" under a food standard established by the federal government —in consultation with industry. The chemicals then become a part of the food's "identity" and are exempt from disclosure to the public. For example, you don't have to be told on the label when brominated vegetable oil is in ice cream, or that caffeine is a required ingredient in cola drinks for them to be called colas, or that about 95 percent of all the caffeine in such drinks is not natural but is added artificially by manufacturers to provide a "lift." Parents who wouldn't think of letting their kids drink coffee let them instead consume comparable amounts of caffeine in soft drinks.

James S. Turner in his book *The Chemical Feast* noted that there are 485 food chemicals and other ingredients which do not have to be listed on labels. The standards are so inconsistent—having been hammered out privately between FDA and industry representatives—that the same chemical may have to be listed as an ingredient of one food and not of another. MSG must be noted on the labels of soups, for example but not on those of mayonnaise or salad dressings. Thus by looking at labels you have no way of knowing whether the additive is not present or is present and simply not listed because of an FDA whim. Sometimes the additive is simply identified as an "emulsifier" or "artificial flavor."[1]

As William Goodrich, former general counsel for the FDA, has pointed out, the FDA shies away from complete labeling of food chemicals because it knows the public is skittish about them. When asked by Congressman H. L. Fountain why the FDA didn't enforce a labeling regulation for nitrite in smoked fish, Goodrich replied that

> we might more alarm the consumer about declaring a list of chemicals that we are under an obligation to assure as safe. . . . I am agreeing with you and the committee that there should be more declaration of the composition of food, bearing in mind, however, that there are some instances, particularly in declaring chemicals, in which we have found the chemical itself to be perfectly safe and technologically or physically useful in approving it, and if you declare it on the label it simply precipitates the correspondence between us and the public asking why we allow this horrible thing in food . . .[2]

Do We Really *Need* All Those Additives?

It's surprising that at the turn of the century—a time of little refrigeration, heat sterilization or freezing, and of slow transportation—when you would have thought we needed additives, at least as preservatives, they were soundly rejected by both government and industry.

H. W. Wiley, chief of the USDA's Bureau of Chemistry, the predecessor of the present FDA, wrote in his annual report for 1908:

> A large number of prominent manufacturers during the year entirely abandoned the use of any kind of preservatives and openly announced their adhesion to the doctrine that drugs should not be placed in foods. Although there have been no suits brought so far involving the addition of chemical preservatives to food, the practice has been so openly discredited by so many first-class manufacturers as to warrant the statement that the cause of pure food, in so far as chemical preservatives are concerned, has been firmly established.[3]

Today, both industry and the government have done an about-face. Though our laws against food additives, by Congressional intent, are much tougher, food chemicals still proliferate. To illustrate how far out of hand things are, the government does not even know how many additives are being used or by whom, or for what; official estimates range from three thousand to ten thousand. Since 1958 the FDA has issued formal regulations approving over three thousand food additives and declared nearly one thousand "generally recognized as safe."

If we didn't need so many chemicals in foods—or "drugs in food," as a conservative industry then accurately phrased it—sixty-five years ago, why do we need them now when our preservation methods are more advanced? What are they? What do they purportedly do?

Some additives are natural, for example spices, such as ginger and nutmeg. Others are imitations of natural substances such as synthetic vitamins. Still others are totally synthetic, invented in a chemist's head and unknown before the twentieth century. Essentially, food additives are used for the following purposes:

Nutrients. These are substances, such as vitamins (niacin, carotene), and minerals (calcium, iron, phosphorus, iodine), put in food to "enrich" or "fortify" it. Many of these nutrients must be put into food because they have been stripped from such staples as bread, rice and potatoes during processing. For example, the USDA found that processing potatoes into dehydrated flakes for "instant mashed potatoes" removed 75 percent of their vitamin C, as compared with a loss of 44 percent in ordinary cooking. Thus, "instant mashed potatoes" must be fortified with fifty milligrams of vitamin C per ounce before they can be sold. For the most part, however, it is important to remember that although the industry points with pride to nutrient additives, few—only 7 percent, according to one calculation—of the additives being used have any nutritive value whatsoever.

Flavors and flavor enhancers. These include cloves, pepper and so forth, as well as synthetic fruit flavors, such as strawberry, walnut, wintergreen, used in place of the real ingredient to give us "that strawberry-flavored food year round." Flavor enhancers, such as monosodium glutamate (MSG), disodium guanylate and inosinate, bring out the flavor in foods (some of which is detroyed in processing) and enable a manufacturer to add smaller amounts of the real ingredients for flavor.

Preservatives and antioxidants. These keep meat from changing color, bread from becoming moldy, fats from tasting and smelling rancid. The best-known antioxidants are BHA (butylated hydroxyanisole) and BHT (butylated hydroxytoluene), which are somewhat capriciously added to food, for certain manufacturers by using better manufacturing practices get along perfectly well without adding them to retard spoilage. Such preservatives are intended to increase a product's shelf life without the expense of refrigeration or the necessity of more frequent shipping to insure fresh products. As one food maker said, "It's the age of ageless food."

Emulsifiers, stabilizers and thickeners. These are the substances that make cream seem thick, keep the oil and vinegar in salad dressings from separating, and generally give a smooth, uniform texture to bread, bakery products, ice cream, puddings, shortenings. As Dr. Michael F. Jacobson points out in his book *Eater's Digest*, some manufacturers use a recipe that automatically produces a food with satisfying texture and consistency. Other manufacturers of the same products rely on the above group of additives to cover up the fact that inferior ingredients or poor manufacturing practices make their product watery, lumpy or crystalline.[4]

Acidulants (acids, alkalies, buffers and neutralizing agents). These are designed to control the acid–alkali balance in such foods as soft drinks, butter, sherbets, jellies and jams, canned vegetables, processed cheese—to keep them from being too tart or too sour. They're used sometimes to protect the color of canned vegetables and to neutralize the acid in such products as tomato soup, but sometimes it's a case of having to use a neutralizer to counteract the acid from another food additive.

Colors. A few are natural, but over 90 percent are synthetic —most of them coal-tar derivatives—and they go into nearly every kind of food and drink imaginable: meats, wines, soft drinks, cakes, gelatin desserts, bread, fruits, cereals, potato chips, to give the food the color the industry believes consumers associate with high quality, whether or not that high quality is actually present in food. You'll note on labels "U.S. certified artificial color" (though they don't tell you exactly what kind of color), and the "U.S. certified" gives the impression it's been certified for safety, which is not correct. "Certified" means simply that it has met certain government standards which guarantee neither safety nor purity, for the standards allow certain percentages of impurities, such as traces of arsenic and other compounds, to be present. An 80 percent "pure" dye, for example, contains impurities of 20 percent, produced during synthesis.

Bleaching and maturing agents. Used to make flour white and less coarse, as well as to bleach other products such as cheeses.

Sequestrants. To inactivate minerals in foods or beverages, most notably the water in carbonated beverages to keep the product from becoming cloudy.

Humectants. To keep moisture in certain foods, for example shredded coconut and marshmallows.

Anticaking agents. To keep salt and powders such as pudding mixes free-flowing.

Firming agents. To keep processed fruits and vegetables from becoming soft.

Clarifying agents. To remove small particles of minerals such as iron and copper which could make liquids cloudy.

Curing agents. For example, nitrite that also stabilizes color and may have some preservative effect.

Foaming agents and foam inhibitors. The first gives us pressure-packed whipped toppings, the second helps keep foam off orange juice.

Nonnutritive sweeteners. Namely saccharin, since cyclamate was banned in 1970.

Perhaps it struck you, as it did us upon reading this list—which was taken from a booklet of the Manufacturing Chemists' Association[5]—that although some additives seem of real value (in preventing spoilage, for example), many are used for strictly cosmetic purposes: to make food more colorful, less cloudy, thicker, smoother, more cohesive; in short, to make it appear artificially what it is not.

In truth we have been conditioned by the food industry to accept consistencies and colors in foods that are foreign to the natural product. If you have tasted old-fashioned peanut butter lately, you know that the coarseness of the pure ground peanuts is a far cry from the smoothness of the modern emulsified product. The thickness in today's dairy cream does not come from the cow; it's artificial, the product of thickeners. Old-fashioned hot dogs were not as red as today's; modern ones are fixed bright red with the chemical nitrite. That gorgeous dark color in the devil's-food cake you buy is achieved not with chocolate but with dyes, so that the maker can get away with putting less chocolate in. So-called "egg bread," more expensive than ordinary white bread, is colored yellow to simulate the yolk; a loaf of such bread has on the average but one half an egg.

Dog owners may be interested to know that dog food is colored with dye, even though dogs are color blind and couldn't care less whether their food is green, purple, black or colorless. A dog does not know that hamburger is red when you give him the real thing. Nevertheless, dog food is tinted a red meat color to please the human buyer, just as baby food is concocted with salt and additives not because they add nutritional value but to please the taste of mothers.

Even Florida oranges must be dyed orange because during

certain seasons they are a spotty green even when ripe, and suffer—so the industry says—a competitive disadvantage. The FDA could put an end to this nonsense by simply telling the public that not every orange in its mature, ripe sweet stage is orange—that there are some varieties of orange that when fully ripe are still green because of soil or weather conditions. But instead the industry dyes the skin; and the dye is then ingested when the skins are eaten, as in marmalade. Low-grade pistachio nuts are colored red; so are some potatoes. Liquors and cordials which used to be made with fruit now contain artificial dyes and flavor.

Food technologists are nearly hysterical in their insistence not only that foods be colored but that they be colored precise shades. According to industry spokesmen, food processors strive for perfection; for example, they dye imitation orange drinks the precise color of fresh orange juice. They insist that slightly off shades tend to be associated in the buyer's mind with deteriorating quality. Of course, the food may in fact *be* deteriorating, and one of the purposes of additives is to mask that deterioration.

It is clear that certain additives—surely colors—are often used deceptively to convey that a product is of higher quality than it is, even nutritionally, as with "egg bread." You believe that somehow you are getting more nourishment when in fact you are only getting more coal-tar yellow dye.

To illustrate how habit and the most minute cosmetic concerns can dominate the use of additives: rice for years has been coated with glucose and talc which contains asbestos. The reason, according to the Rice Millers' Association, which opposes the practice, is that the coating gives rice a glossy appearance, which presumably consumers find more appealing than the natural dull look. The coating, however, the millers note, "adds nothing to the quality, nutrition or cooking characteristics of the rice," and is used simply because of "long-standing habit." In fact, there are a number of drawbacks: the shiny coating conceals the true appearance and nature of the rice kernel, adds cost, and necessitates washing that causes a loss of vitamins and minerals.[6] Moreover, asbestos produces cancer of the esophagus, the stomach, the colon and the rectum—as workers in asbestos plants have unhappily discovered.

Because knowledge about the technical uses of food additives is so tightly controlled by industry, it is difficult for consumers to determine which chemicals are what we would call "necessary." But there is a growing realization that even for current accepted usage there is a superfluity of food chemicals. Many are arbitrarily used, perhaps because of outmoded manufacturing practices, out of habit or to cover up inferior products. Some are redundantly used.

Take, for example, food flavorings and dyes. A scientist at the Department of Agriculture has estimated that of the seven hundred chemicals now used for flavoring, about thirty could accomplish the same thing.[7] (France, that country of gastronomical excellence, uses seven.) A chemist has pointed out that printers get along with three primary colors—red, blue and yellow—from which they produce every nuance of color imaginable. The food industry, however, must have its own special yellows, oranges, greens, blues, magentas, violets for every occasion.

Even in preservation, the area where most of us would consider additives useful, there's doubt that all are really necessary. Dr. Jacobson noted in testimony in 1972 that some makers of vegetable oils, potato chips, shortening and peanuts add BHA and BHT (preservatives) to their products, while their competitors do not. "For instance," he said,

> Wesson (soy and cotton) oil does not contain any preservatives, whereas Safeway's Nu-Made (soy and cotton) oil contains BHA and BHT. Crisco solid shortening does not contain a preservative, whereas Spry contains both BHA and BHT. Jay's and Wise potato chips do not contain preservatives, but Lay's and Sunshine potato chips contain both BHA and BHT. Red Star dry yeast does not contain a preservative, but Fleischmann's dry yeast contains BHA. Planter's peanuts do not contain a preservative, but Fisher's peanuts contain BHA. I could go on and on. It is clear that BHA and BHT are very frequently used unnecessarily.

Although the FDA considers these antioxidants "generally recognized as safe," Dr. Jacobson advises consumers to avoid them (they are listed on labels) because they have never been tested for cancer induction and have showed some signs of harm in animal studies: increase of cholesterol in blood, loss of hair, and birth defects.

The following table, from Dr. Jacobson's 1972 testimony,[8] lists some brand-name products which do and do not use antioxidants.

Product Category	Antioxidant(s)
shortening (liquid)	
Wesson (soy and cotton oil)	—
Wesson (soy oil, buttery flavor)	—
Nu-made (soy and cotton oil) (Safeway)	BHA, BHT
Crisco (soy oil)	BHA, BHT
shortening (solid)	
Crisco (Procter & Gamble)	—
Spry (Lever Brothers)	BHA, BHT
Velkay (Safeway)	"oxygen interceptor"
nuts	
Planter's Spanish peanuts	—
Planter's mixed nuts	—
Planter's peanuts	—
Fisher's peanuts (Beatrice Foods)	BHA
Spanish peanuts (Giant Stores)	BHA
mixed nuts (Giant Stores)	BHA
Virginia peanuts (Giant Stores)	BHA
potato chips	
Jay's	—
Wise	—
Lays	BHA, BHT
Sunshine	BHA, BHT
Utz	"oil preserver added"
Snyder's of Hanover	"antioxidant added"
pork sausage	
Luter	—
Gwaltney	—
Super-Right Country Treat (A&P)	—
Super-Right Old-Fashioned Farm Style	BHA, BHT
Park's Little Link Sausages	BHA, BHT
canned puddings	
Delmonte	—
Filberts	—
Hunt's Snack Packs	BHA, BHT
Jello Pudding Treats	BHA

Product Category	Antioxidant(s)
toaster tarts	
Nabisco Toastette	—
Merico Jelly Jump Ups	—
General Foods Toastem	BHA
Danka Toastem	BHA
Kellogg's Pop Tarts	BHA, BHT
frozen French toast	
Downyflake	—
Aunt Jemima	BHA, BHT
instant potatoes	
French's potatoes au gratin	—
French's scalloped potatoes	—
Betty Crocker potatoes au gratin	BHA
Betty Crocker scalloped potatoes	BHA
gelatin dessert	
Jell-well lemon flavor (Safeway Stores)	—
Jell-O lemon flavor (General Foods)	BHA
dry yeast	
Red Star (Universal Foods)	—
Fleischmann's	BHA
peanut bar	
Planter's Peanut Block	—
Snicker's Peanut Munch (Mars, Inc.)	BHA, BHT
peanut-butter-cups candy	
Reese	—
Boyer	BHA
dry soup mixes	
Lipton's green pea	—
Lipton's chicken-rice	—
Lipton's tomato-vegetable	—
Lipton's Cup-a-Soup green pea	BHA
Wyler's chicken-flavored rice	"oxygen interceptor"
Lipton's Cup-a-Soup tomato	BHA
liquid pop mixes	
Otter pops	—
Cherri Aid	BHA
Kool Pops	BHA
desserts	
Royal No-Bake cheesecake	—
Jell-O cheesecake	BHA, BHT

There's no better proof of the needlessness of some chemicals than the alacrity with which foodmakers have been able to jump in and make additive-free food in response to consumer demand. *Fortune* reported:

> Some food companies have found a good thing in the additive scare. The "natural," "organic" and "health" food businesses are growing at a tremendous rate: according to some projections, sales in health-food stores and other retail outlets may reach $550 million this year [1972], up from $400 million last year. . . . The Jones Dairy Farm offers sausages without "unnatural" preservatives; Dannon Milk Products boasts of yogurt with no chemical additives. Borden's Sacramento Foods division is test-marketing organically grown tomato juice.[9]

Further, Quaker Oats has come out with a new additive-free cereal "made by Nature, not by man." Some manufacturers have found alternatives for brominated vegetable oil—found to cause heart lesions in animals—though others continue to use it. Certain beers, Meister Brau, for example, contain additives to stabilize the foam; other brewers, such as Budweiser and Rheingold, say they use no additives at all.

If some companies can get along without additives, why can't others?

For Whose Benefit—Theirs or Ours?

Given a choice—which we can exercise only through political power—would we really want all those additives in our food? How much do the chemicals benefit us? And how much do they benefit the foodmakers? If we had to take a dangerous drug, we would want to know its potential benefit compared with the risk. But with food additives we often take risks for little or nothing. The benefit–risk ratio for food chemicals often boils down to a great big health risk to you and a great big economic benefit to industry.

Industry spokesmen sometimes hint that without additives we might starve to death, that malnutrition would sweep the world. This, they say then, is the price we would have to pay for forsaking food additives. Not only is this a bit of an overstatement, but it warps the argument. Few sensible persons suggest that all additives should be summarily banned—we say only that those used should serve some useful purpose and be examined thoroughly for safety. This standard seems reasonable to most people, except perhaps to certain commercial groups which have built an empire on additives and now fear that any shaking of faith in their chemical money tree could cause the whole lucrative enterprise to crumble.

Instead of judging additive safety on its own merits, industry insists that we should take "calculated risks" in additive use. That is, we should trade off what they call "benefits" against "risks." *Fortune* said: "Although not many consumer advocates will acknowledge the fact, additives have benefits as well as risks, and any reasonable policy should be based on a weighing of the two." Certainly some additives have benefits, although they are ill-defined. Probably some do protect our health, save us time and money, prevent wasteful spoilage and give us "convenience" foods which we could not otherwise have. On the other hand, it seems clear that many are primarily cosmetic, substitutes for good manufacturing, shipping and storage practices, and a shortcut to increased profits for industry.

Moreover, when industry spokesmen toss around the term "benefit–risk," what do they really mean? Do they mean consumer health benefits weighed against consumer health risks? Or consumer economic benefit against consumer health risk? Or some kind of consumer social benefit (such as time-saving) against consumer health risk? Or, on the other hand, do they mean *industry economic benefit* against *consumer health risk*?

Whose welfare is industry really considering in this "benefit–risk" ratio which, they say, we should take cognizance of in evaluating food chemicals?

All too frequently, the ratio is a weighing of private profits against risks to society at large. Admittedly, industry policymakers may share in the personal-health risk, but this is often lost sight of or rationalized away in the drive for a "maximization" of corporate profits.

In short, industry is asking us to trade the possibility of long-term risk to *our* health in return for shortsighted, immediate economic gain for a small segment of society. Furthermore, in case the scheme should backfire and health damage result, they ask us to bear the economic cost. As

Dr. Samuel Epstein, cancer researcher and professor at Case Western Reserve Medical School, has pointed out, one malformed child, possibly the product of the irresponsible use of additives, can cost the state $250,000 for medical care alone. When the damage is widespread the social costs are inestimable.

Before you can understand the industry's sensitivity to more widespread knowledge about *consumer* benefits and hazards of food chemicals, you must recognize its enormous financial stake in the matter. There is no question that the increased use of chemicals has brought a bonanza to the food industry. The chemical companies and the foodmakers (which are fast becoming indistinguishable) are interlocked in a mutually profitable venture with no end in sight. "As the food industry grows, so grows Pfizer" (Chemical Company), as a recent institutional ad put it.

The food industry is the largest and fastest-growing in this country. Its sales in 1971 were $139.2 billion, up 63 percent since 1960. Trade journals attribute the growth to new convenience foods, created wholly or partly out of additives. Remember, synthetic chemicals are much cheaper substitutes for flavors and colors than real fruits and vegetables, and the profit on synthetic foods formulated from chemicals is enormous. With these two facts in mind, the large chemical companies, looking for areas in which to expand in the early 1960s, chose food because of its great promise, and they haven't been disappointed. The president of Monsanto has said, "Our food ingredients program [a euphemism for food chemicals] is rapidly fulfilling its early promise. We expect its growth and profitability to accelerate as we become a closer working partner in the food-processing industry's efforts to improve the flavor and convenience of its products."[10]

One has only to look at the trade journals of the food industry to note the omnipresent hand of the chemical industry in our food supply. One ad reads: "Food has a better future with Pfizer" (Pfizer, Inc., Chemicals Division). And another tells us that Durkee has "technology that tops nature"; the company specializes in "recreating the pure, special individual flavors of popular foods" through science. We discover in the February 1973 issue of *Food Engineering* that Pfizer has a new powerful artificial flavor, called methyl cyclopentenolone. It has a "nutty odor" and the "taste of maple or walnut." It's recommended for ice cream, candy, baked goods, desserts, chewing gum, beverages, syrups, and it's guaranteed to save food processors money because it's so powerful that you use less of it—and, of course, you don't have to use old-fashioned real walnuts or maple to get the flavor.

From the February 1973 issue of *Food Processing* comes news that food processors can "reduce their requirements for tomato solids: by using an array of synthetic tomato flavors, and save a bundle of money, too." Touting the products, the article states: "As a flavor that can replace up to 25 per cent of the tomato solids in a formulation, Red Tomato Flavor 567 is recommended to standarize the flavor profile of tomato containing products. Four ounces of the ingredient replaces the flavor of 100 pound tomato solids." And another: "Cooked, ripe tomato flavor offered in either dry or liquid form . . . is recommended for use in soups, sauces, dips, salad dressings and convenience foods . . . 1 pound replaces the flavor and aroma of 1200–1600 pound tomato juice at a cost of $5.00 . . ."

In 1955 an estimated 419 million pounds of additives went into our food. Today the figure is 1.06 billion pounds annually —or five pounds per capita. Every year $500 million worth of food chemicals are sold, which may not sound overwhelming, but when the chemicals are distributed in food, the economic importance is magnified into a hundred-billion-dollar enterprise. Thus, understandably, it is not only chemical companies but food magnates generally who get queasy stomachs when additives are criticized. Not only do they stand to lose money, but they may have to reformulate products to achieve a new agreeable mixture of chemicals.

Nor are we at a plateau, as more and more ideas for synthetic foods spring full-blown from the chemist's brow. Almost any "food" imaginable can now be fashioned by the dexterous mixing of artificial chemicals. The proliferation of these foods contributes to the burden from additives. In the early 1960s there were about 1,500 grocery items; today there are some 32,000—about 5,000 new ones are added yearly.

Even though additives produced in great quantities by

chemical companies are relatively cheap, their addition to food often raises prices and profit margins. Observed *Chemical and Engineering News* magazine: "Food additive companies find that it is easiest to sell a new additive to the food industry if it reduces significantly the cost of making the food or decreases the production time." The hawking of additives is based primarily on claims that they make production cheaper. Some advertisements from trade journals:

> Nestlé Vee-Kreme makes cream old-fashioned. Wherever you apply fresh dairy cream, replace it with Nestlé Vee-Kreme. You will discover that Nestlé Vee-Kreme successfully replaces a whole series of other ingredients in your formula. This saves you time, trouble . . . and money!

> Is your flavor thinking old-fashioned? If you are locked-in on flavor enhancement, "Ribotide" is the best way to break out, improve flavor, avoid problems. And save substantial sums of money, too.

> A fabulous new food . . . TVP [textured vegetable protein] could hardly look or taste better . . . or be more economical. It is available in granular, chunk, dice, strip and chip forms. It comes unseasoned, or with flavoring of almost any kind—meaty, nutty, tangy, salty, even fruit flavors. . . . It's an excellent enrichment for casseroles, snacks, stews, gravies, ground meats and many convenience foods. Find out more about this fabulous new food . . . about the profit making opportunities it affords.

This does not mean that when they take the real cream out and put in a cheaper substitute, the price is lower for us. On the contrary, additives usually result in higher prices, even when they are much needed nutrients to replace those which have been processed out in the first place. The cereal Total, according to Dr. Jacobson, was created by coating Wheaties with vitamins. The company, he says, added a half cent's worth of vitamins to twelve ounces of cereal and then raised the price by eighteen cents—to forty-five cents. A package of powdered eggs, the equivalent of two eggs, costs thirty cents, or about three times what you would pay for two real eggs at five cents each.

Sidney Margolius in his book *The Great American Food*

Hoax analyzes example after example of how much more the new synthetic and convenience foods cost per unit compared with conventional foods. For example, he notes, General Foods' Cool 'n Creamy pudding costs forty-five cents for four servings, compared with about thirty-two cents for the same amount you could make from a pudding mix combined with your own milk.[11] Color also is a prime regulator of price. Cherry pop is merely water, sugar, flavorings and dye. But who would pay twenty cents for a bottle of colorless liquid? It is, as Anita Johnson of Ralph Nader's Health Research Group points out, the color that commands that price.

Whenever food processors speak of using additives, they say that the additives "improve" the product. What they mean by "improve" is vague—perhaps they mean improve the food in uniformity, gimmick dispensers, product acceptance, sales profiles. It's not likely they mean improve in terms of nutritional value. Actually, additives can harm the taste, texture and nutritional value of food. Some of the new artificial chemically created foods contain no protein, minerals or vitamins. Composed exclusively of additives, carbohydrates, fat and water, they are empty synthetic food providing empty calories just as alcohol does and do not fit the classic definition of food as nourishment.

There is also evidence that the eutrophic (nourishing) value of these processed foods or those "put together from ingredients other than those derived directly from plants or animals" is less than that of natural foods, according to Roger J. Williams of the Clayton Foundation Biochemical Institute at the University of Texas, who is participating in a study on the subject for the National Academy of Sciences. He says that many snack foods—such as crackers, candy bars, potato chips, cookies, soft drinks—have "very low or even zero eutrophic value," which would "certainly result in underdevelopment."[12]

It is ironic, then, that our new "convenience" foods are giving us far fewer nutrients though we are paying more for the food itself. But perhaps it's all in the way you look at food. When asked why society should tolerate any potentially dangerous additives at all in food, industry spokesman Emery C. Swanson of Swanson and Associates, which develops new food products, replied: "We are a hedonist society. People don't eat for nutritional purposes."

Nor can we console ourselves that the food tastes better. Many authorities have long commented on the deteriorating quality of many common foods, which is also apparent to any consumer over the age of twenty. For example, until recently when DES was banned in feed, 80 percent of the beef cattle in the country was raised on DES feed to promote growth. And this was supposed to be of great benefit to us as consumers, according to the industry. But Dr. Gordon W. Newell of Stanford's Research Institute has pointed out that DES "gives a silky, watery texture" and also makes the beef curl and shrink excessively during cooking, with a loss of flavor.

Despite rare consumer victories in which dangerous additives have been banned, we face a future of not fewer but more food chemicals. Industry representatives predict that by 1980 the sales of food additives will reach $765 million—a 50 percent jump from the current $500 million per year. They say the use of flavoring materials will be up 50 percent; stabilizers, 50 percent; surfactants, 40 percent; flavor enhancers and potentiators, 100 percent; acidulants, 60 percent; synthetic sweeteners and bitter agents, 60 percent; antioxidants, 100 percent; preservatives, 67 percent.[13] This outlook for a bigger, better "chemical feast" might better be described as a collective Last Supper. For we have virtually no idea of the cumulative chemical effect on living tissue, especially after years of consumption.

"But They're only Chemicals."

Some industry representatives prefer the term "food ingredient" to "chemicals" or "additives" because they know the public's adverse reaction to chemicals in food, and they inform us that chemicals are, after all, the building blocks of the universe and are not to be feared *per se*.

Says the Manufacturing Chemists' Association:

> Too often people without scientific background believe that the word "chemical" means something dangerous or, at best, "unnatural." They fail to realize that everything in the world is

essentially chemical—from the concrete and steel in the Empire State Building to the vitamins in our food. The average homemaker would look incredulous if she were told she just fed her family triglyceride esters of palmitic, oleic, linoleic, and stearic acids. But when she becomes aware that the words are simply chemists' terms for the chief components of cooking fats and shortenings, she begins to understand that foods are made up of many chemicals.

Hartley W. Howard, technical director of Borden, Inc., has said, "Consumers must learn that 'chemical' is not a dirty word; that our very bodies consist of nothing but chemicals—maintained and replenished by the chemicals in the foods we eat, the chemicals in the water we drink, and the chemicals in the air we breathe. Once consumers learn that those substances with the unfamiliar names are but a small addition to the vast number of chemicals provided by the more familiar food ingredients we will have made a considerable step forward."

Officials at the FDA have been heard to mimic the same sentiments in defending food additives, hoping the public will get the message that chemical food additives are good for us. It is an absurd argument to deduce that because some chemicals are good for us, therefore all are. What about arsenic, for a start? Clearly there are beneficial and harmful chemicals, even "natural" ones.

Furthermore, the advocates of chemical additives *per se* do not point out that many of them are not "natural" chemicals but man-made creations which nevertheless have potential biological activity when taken into the body. Through only minor alterations of a chemical structure—the addition or subtraction of an atom, for example—a chemist can come up with an entirely new substance, previously unknown, with unique properties which will react in an undetermined manner.

Perhaps more sobering is that we are deviating farther and farther from the natural configurations of chemicals in our quest for new artificial ones. What these chemicals do biologically, nobody knows.

Besides the additives intentionally added to foods, there are indirect, unintentional additives and contaminants to contend with. These seep in accidentally—such as additives intended only for packaging or other industrial use, pesticide and antibiotic residues, and dangerous contaminants such as poison-producing molds.

Undoubtedly some additives are beneficial, improve the quality of food and can be shown to be harmless at this stage of science. The removal of *all* additives would probably, as one expert has said, cause "chaos in our food supply," and would be totally needless. But we can trim the use of additives to those which have a positive value, are necessary and not just frivolously used, and, of course, are indicated to be safe.

Former FDA attorney Goodrich testified in 1971, "As the generally-recognized-as-safe list became a popular subject of discussion last summer or thereabouts, people began to ask us why is it that certain things are in foods. Sometimes we have a good reason; sometimes not."

There should be good reason. Because additives have gained such a stronghold in our lives, it is time to take a sensible look at their use and stop to consider what they are doing *for* us compared with what they are doing *to* us.

Fast Poison, Slow Poison

All of us are involved in a gigantic experiment of which we shall never know the outcome—at least in our lifetime. How dangerous are the food chemicals we are eating? Are they contributing to cancer? To birth defects? To mutations? To liver, brain and heart damage, and to a hundred other diseases? We don't know. We can only guess in some instances. And in others it is a blind experiment of massive proportions.

Our knowledge of which foods are poisonous and which are not is largely based on accidental information handed down for centuries. For years people didn't eat tomatoes because the tomato plant is a member of the nightshade family, whose berries are highly toxic and can cause stupefaction and paralysis, sometimes culminating in death. We don't go out gathering a skirtful of castor beans or buckeyes or certain kinds of mushrooms for dinner or cook up some hemlock roots (which resemble wild carrots) or hemlock seeds (which can be mistaken for anise) for precisely the same reason. Common knowledge based on collective observation tells us that such consumption is quickly followed by vomiting, convulsions, delirium, weakness, paralysis and sometimes death.

That we prefer some foods and shun others is merely a reflection of our concept of *acute*, or rapid, poisoning. We generally regard as "safe" those foods that don't incapacitate or kill us within a period of time that can be directly linked to their consumption. From the earliest time, our highest toxicological standard was acute poisoning, and this was the extent of our concern about anything we took into our mouths until perhaps only the last fifty years.

When a few food additives were tested at the turn of the century, the procedure was for a scientist to feed the chemical to a few dogs or rabbits, wait a few hours or days or perhaps as long as a week, and note any sign of diarrhea, nervousness, disorientation, vomiting or other classic symptoms of poisoning of the type associated with a known acute poison like arsenic. If the animals survived or didn't show untoward effects, the additive was declared safe and went into our food. A primitive test indeed—but many of the food additives we eat today have never been tested in any other way. All we know is that if we eat them we aren't expected to die—at least within twenty-four hours or so.

But toxicology has become much more sophisticated in the last few years. We now realize that far more critical than acute poisoning is the subtle, long-term insidious poisoning of the body by certain chemicals that work slowly or cumulatively and whose ravages may not become evident for many years. There's cancer, and brain damage, and heart, liver, and kidney and blood diseases, and birth defects, miscarriages, and mutations, and no one knows how many other bodily disturbances that can be associated with the intake of environmental chemicals over a prolonged period or at a critical stage of pregnancy. We know that birth defects are not the result of sin, but often of accidental contact with viruses and chemical pollutants, such as additives and pesticides. We know that chemicals may leave their mark not only on the recipient but on future generations, through mutations which might not be detected for a hundred years or more—if at all.

Geneticists Joshua Lederberg at Stanford and Bruce Ames at the University of California are both worried that our gene pool now is being poisoned by the careless use of additives. Dr. Lederberg points out that the first chemical mutagen ever discovered, in 1944, which caused a "small landmark in the history of genetics," was a food substance—oil of mustard. It caused genetic damage in fruit flies and launched a whole new study of mankind. Since that time, organic peroxide, widely used for bleaching starch and maturing flour, has been declared a mutagen. So has bisulfite, used extensively as a food preservative, and cyclamate. In 1968, Dr. Marvin Legator, geneticist and chief of the cell biology branch at the FDA, produced "impeccable work" (Dr. Lederberg's assessment) showing that cyclamate's metabolic breakdown product produces chromosome breakage in rats. On that basis alone, cyclamate should have been immediately banned, says Lederberg. But the FDA completely ignored Dr. Legator's work.

It is important to note that such chemical mutations are almost always harmful, not beneficial. In some cases the impact from mutations might be immediate, causing sterility, miscarriage, eye tumors, extra fingers and toes, dwarfism. According to Dr. Victor McKusick, geneticist at Johns Hopkins, certain types of dwarfs are invariably the result of *new* mutations in the genes of normal-sized parents. Or over the long term, chemical damage can cause hemophilia, Mongoloidism and other mental retardation as well as more subtle, imperceptible harm such as a lowering of intelligence, loss of physical vigor, a susceptibility to disease, more rapid aging—in sum, a "general reduction in the viability of succeeding generations similar to what we would expect from radioactive fallout," in Dr. Lederberg's words.[1]

The worst aspect of mutations is that they change the human blueprint, becoming an irreversible legacy passed from generation to generation. Even one mutation in a million people is a serious problem, says Ames. With each new branch of the family tree, the damage is magnified, passed along from the original donor to his children, his grandchildren, his great-grandchildren and their great-grandchildren. As Ames warns, "If we're filling ourselves now with mutant genes, they're going to be around for generations and generations." Moreover, in all likelihood the damage would be so diffuse and gradual that the cause–effect relationship would go unnoticed by the human race.[2]

Birth defects, as distinct from mutations, are not passed along to progeny, although they too are irreversible and may remain well hidden. Science had known for several decades that chemicals could produce deformed children, but it was not until the thalidomide tragedy that the problem became a public issue. Since then drugs must be meticulously screened for their potential action on embryos; however, as noted previously, food additives with similar potential hazard are put into another category for unfathomable reasons. For there is overwhelming evidence that food chemicals can cause readily noticeable structural defects in the newborn, such as twisted spines, shortened limbs, incomplete skulls, absence of eyes, cleft palate, web feet. The study of such defects is called teratology (from the Greek word *teras*, meaning monster), the study of monstrosities, and the chemicals causing such destruction are called teratogens.

The readily observed deformity caused by chemicals may be only the tip of the proverbial iceberg. More subtle chemically caused deformities may also be contained in the new body: minute internal damage discoverable only under a microscope, or never-to-be-discovered functional disorders, such as a faulty pituitary gland which can wreak all kinds of bodily havoc, brain damage resulting in mental retardation, a predisposition to diabetes or kidney disease or cancer, a subtle heart defect, central nervous system problems, learning disabilities. All are possible deformities that can be passed on to children by mothers who happen to be vulnerable to the onslaught of a particular chemical at a specific time of pregnancy.

A University of Michigan team of investigators led by Dr. Joan M. Spyker recently found that mice injected with methyl-mercury dicyanimide (which may be present in fish, flour, sugar, dairy products, meat and potatoes at low levels) produced offspring that visibly seemed fine—even when their brains were examined microscopically no neurologic damage could be spotted—but their behavior was bizarre. Unlike control animals whose mothers were given no mercury, the baby mice from mercury-treated mothers often refused to groom or explore; half of them walked backwards and when put into water thrashed around and sank. Thus, as the investigators observed, severe birth damage from chemicals may go undiscerned when it is manifested in subtle behavioral dysfunction instead of in physical appearance.[3]

It's likely that if thalidomide had caused mental retardation instead of the gross skeletal deformities it did, we would still be using it today, dispensing it like aspirin.

Furthermore, birth defects may show up only later in life. In the early 1950s diethylstilbestrol (DES), the cattle hormone, was given as a drug to pregnant women to prevent miscarriages. It was not until fifteen or more years later, when the youngsters were in their teens, that vaginal cancer began showing up in the daughters of the women who had taken DES.

Chemicals can also cause spontaneous abortions. If the damage to the cells is too great, the fetus simply can't survive and miscarriage occurs. Approximately one out of every four pregnancies ends in either a miscarriage or a deformed child. According to the National Foundation/March of Dimes, 250,000 children are born every year with defects, both observable and unobservable at birth. The Foundation estimates that environmental factors, including chemicals such as food additives, contribute to at least 20 percent of the abnormalities. (Another 20 percent is probably due to hereditary factors and the remaining 60 percent to factors as yet unidentified.)

Ever since Percival Pott in 1775 linked soot to the scrotal cancer of chimney sweeps in England, scientists have had inklings that chemicals have great bearing on the existence of cancer. Since then it's been well established that chemicals, including aromatic amines and asbestos, induce cancer in people in other occupations. Of course, the most dramatic evidence comes from the reputed legacy of Sir Walter Raleigh: tobacco. In 1971, some 335,000 Americans died of cancer, 68,000 of them from lung cancer, mostly attributable to smoking.

Cancer may also relate to genetics, possibly viruses, and other individual habits or traits, but scientists are convinced that environmental chemicals are a major factor. Certain kinds of cancer are more prevalent in specific geographical locations. For example, stomach cancer is extraordinarily common in Japan. But when Japanese move to the United States, their cancer rate drops. Scientists therefore reason that the cause has to be in large part environmental. Cancer authorities now link chemicals to cancers of the stomach, lungs, breast, skin, bladder, pancreas, liver, blood. Such is the preponderance of evidence that researchers can say authoritatively that the majority of human cancers are environmentally induced. A 1965 World Health Organization committee agreed with that appraisal.

We have introduced countless synthetic chemicals into our environment, most of them in the last quarter of a century. That the sky has not fallen should not reassure us, since the harm that may result, most notably cancer, can have a long incubation period. The most insidious aspect of such cancer—which often goes unappreciated—is that it takes so long to show up after the subject's initial contact with the chemical. Explains Dr. Umberto Saffiotti, associate scientific director for carcinogenesis, National Cancer Institute: "Cancers develop in man and animals long after the causative agent has been in contact with its target tissue and then disappeared. The time interval (latency period) can be as long as ten, fifteen, twenty years or more." Reported bladder cancers from dyestuffs in England, he goes on to point out, didn't strike some workers until thirty years after they had been exposed, and then they had been exposed for only four years.

Preliminary evidence released by the National Center for Health Statistics in April 1973 indeed indicates that we may be starting to see a rise in the cancer toll due to the boom in chemical usage following World War II. The Center reported that in 1972 the cancer death rate rose at the fastest pace in twenty-two years—from 161.4 per 100,000 persons in 1971 to 166.8 in 1972. Experts reviewing the data agreed that increased exposure to chemical carcinogens was involved in the rise.

We could at this moment be sowing seeds for a cancer epidemic in the 1980s or 1990s. Dr. Saffiotti concurs: "The fact that it may take twenty years to detect in man the cancers due to the exposure of a new chemical carcinogen means that the chemical can be given to people for twenty years under the false assumption of harmlessness. If the effect is then detected and properly attributed to the specific chemical and this is then removed from the environment, the cancers it induced continue to appear for the next twenty or thirty years." Using thalidomide again as an example, he speculates that if the sedative had caused cancer instead of birth defects, the evidence of harm would not yet be apparent, though "a large number of people would have been born with a built-in sentence to early death by cancer."[4]

If the cancer shows up in ten or fifteen years, this is a sign that the carcinogen is pretty potent; but if it's weak, it could take as long as thirty years. Also, effects of cancer agents are cumulative (small doses of weak carcinogens as well as larger doses of strong ones can produce tumors) and do not always add up arithmetically, but are synergistic: together they have a potency greater than the sum of their individual potencies. Moreover, even noncancerous substances can potentiate cancerous ones. Aflatoxin, for instance, becomes a hundred times more carcinogenic when combined with hibiscus oil.

Cancer is a special kind of chronic, or long-term, poisoning. But, naturally, chemicals can cause other kinds of damage to tissues and vital organs, even though, as in cancer, the causative mechanism is not understood. In man, a prime example is alcohol, which used over a period of time can lead to cirrhosis of the liver—a hideous disease. High blood pressure is associated with a high intake of the common chemical sodium chloride, or table salt. New evidence shows that an excess of sugar may be linked to heart trouble.

In rats, a preservative, benzoic acid, was found to interfere with growth and in high doses to cause neurological disorders. Monosodium glutamate (MSG) when fed to infant mice destroyed nerve cells in the hypothalamus of the brain. The sequestrant calcium disodium-EDTA (widely used) has caused liver lesions in young rats and kidney damage in humans—the latter when used medicinally to counter the effects of lead poisoning. Carrageenan, a thickener derived from seaweed, has produced ulcers in several species of animals. The flavoring ammoniated glycyrrhizin (licorice) has caused heart failure, hypertension, fatigue and edema in humans who consumed excessive amounts of food containing it. Brominated vegetable oils, used especially in carbonated beverages, has induced heart, liver, thyroid, testicle and kidney damage or changes in rats. Other chemicals have caused diarrhea, destroyed certain vitamins, such as B-12, interfered with vitamin absorption, produced vascular changes, interacted with other substances in the body to cause damage, and generally produced the gamut of biological reactions. There is probably no function of the body that chemicals cannot affect.

Astute medical investigators have traced allergic reactions to a number of food additives—for example calcium and sodium propionates, preservatives to inhibit mold and rope in bread, processed cheeses, cake and other baked goods. As Beatrice Trum Hunter, in her *Fact Book on Food Additives and Your Health*, points out, these food additives are also common ingredients in powders, solutions and ointments used to treat athlete's foot. Ms. Hunter noted: "A physician has reported allergic reactions from propionates, disturbances that began in the upper gastrointestinal tract, four to eighteen hours after eating food containing propionates, and ended with partial or total migraine headaches. Since the gastrointestinal distress symptoms are similar to a gall-bladder attack, the symptoms may be especially severe in cases where an allergy to the propionate is combined with a gall-bladder ailment."[5]

Dr. Jacobson presented another startling case to senators asking about the possible side effects of additives. The case history, here quoted, came to Dr. Jacobson from a physician and involved the preservative BHT.

Female, age 54 years. This patient was on a rather restricted soft diet because of a sore mouth and difficulty resulting from having considerable dental work done. She discovered packaged dehydrated potatoes which she liked very much, and had been eating a great deal of them. On this particular day she had mashed potatoes and tea for lunch. After dressing and driving to the dentist's office she was waiting only a short time when she noted tingling in her face and hands, and she developed extreme weakness and fatigue. This was followed closely by swelling of her lips, tongue, tightness in her chest, and difficulty in breathing. The receptionist summoned the dentist, who administered 0.5 mls. of Epinephrine HCl 1:1000 subcutaneously. Her symptoms subsided in approximately thirty minutes. There was no history of previous drug ingestion and it had been several days since she had any dental work performed, at which time Carbocaine was used with no ill effects.

No further difficulty was experienced until about six months later. This time about fifteen minutes after eating a breakfast of corn flakes and milk she noted the onset of extreme weakness and fatigue and tingling in her face and hands. Remembering the previous episode, she called her husband, who was a physician. He rushed home from a nearby hospital, arriving in about ten minutes, and gave her 0.5 mls. of 1:1000 Epinephrine HCl subcutaneously. Her face was noted to be extremely edematous. The eyes were completely closed, her tongue swollen and she was unable to speak. She also was having difficulty breathing, which was thought to be due to laryngeal edema. Again she responded rapidly to the Epinephrine.

Skin testing the next day by an allergist for corn, milk, potatoes and tea was negative. The only common denominator detected was the butylated hydroxytoluene (BHT) which was present in the corn flakes and potatoes . . .[6]

This, then, is the type of poisoning possibility that we must contend with today: chronic poisoning from a host of diseases, allergies, irreversible cancer, birth defects, miscarriages, mutations—chemical damage undreamt of a short time ago.

Our ability to define the hazard parallels our ability to detect it. We can't talk about what we don't know. Toxicology is still a primitive science, but it is advancing fast, and that is why new concerns about food additives keep popping up. Some people raise the logical question: if these additives, some of which are natural foods, have been used for years, even centuries, why is there no previous evidence that they were dangerous? Isn't this newfound hazard just a quirk of some radical scientist wanting to deprive us of something we've been eating safely for years? Such was some of the public reaction to the banning of cyclamate in 1969.

The answer is that previously we didn't have today's more sophisticated techniques to ferret out danger. However, we may be still overlooking hazards that will seem perfectly apparent to us tomorrow as the science of toxicology advances. One reason much current attention is paid to chemically produced cancer is that our tests for detecting it seem quite reliable. It's entirely possible that chronic use of chemicals causes lots of other hazards, such as immunological impairment or psychological disorders, but at the moment we don't have the scientific know-how to establish a cause–effect relationship.

Testing on Humans or Animals?

How can we detect this slow, pernicious poisoning? Theoretically two ways: observe what the chemical does to humans or observe what it does to animals. Now, there's some difficulty with the first method, because few people want to lend their bodies knowingly to a governmental experiment on the effects of unknown chemicals, though of course that is what all of us are doing involuntarily. If the FDA were indeed a licensed physician, it would probably be hauled up before a group of peers for censure and revocation of license for unethical experimentation with food additives. The case is different with drugs, for it is assumed that already sick persons will consent to an experiment, because they stand to gain substantially if the drug works. No such claim can be made for food chemicals.

Barring direct human experimentation under controlled conditions, the next best course is to try to link up a certain physical damage with a certain additive through epidemiological studies. By careful investigation scientists try to establish a cause–effect relationship of a specific disease or defect to a certain chemical. It was epidemiological data that alerted officials to the dangers of cigarette smoking and thalidomide. Of course, we Americans had been smoking for years before we recognized the danger. And even though the defects from thalidomide were grossly noticeable and were pertinent to a small identifiable segment of the population—pregnant women who had been prescribed the drug—it took *five years* for medical detectives to track down the connection through their offspring.

This will give a clue to how difficult it is to come up with "human evidence" incriminating food additives. Routinely, you will hear the FDA and industry parrot the inanity that such and such a chemical "has never been shown to cause any damage in humans." *Fortune* ("Hysteria About Food Additives") thought enough of this argument to italicize it. To quote: "At this point, quite a few additives have been banned because of questions about their safety; yet *there is no known case of any additive, used properly in a normal diet, having caused any illness other than the kind of allergenic reactions that many foods can cause.*" Gibberish. Such evidence has not been uncovered because ferreting it out is virtually impossible.

Consider: You're eating the same food chemicals as your neighbor, as children in New Jersey, as pregnant women in Utah, as fruit harvesters in California, as steelworkers in Gary, as cigarette smokers in Dallas, as babies in St. Louis. Thousands of chemicals. At the same time you're taking in chemicals from other environmental sources: polluted air and water, possibly cigarette smoke, drugs, cosmetics, pesticides, occupational chemicals, household chemicals. It would take more than the world's super Sherlock Holmes to make any connection between a single chemical in your life and a specific disease or bodily insult. The chemicals are simply too all-encompassing and the diseases too nonspecific and prevalent for a hookup to be made except in the rarest instances.

Another important drawback to human evidence is that damage shows up only after the fact, when we have a legacy of deformed children or cancerous adults. Yet both industry and FDA officials insist that we should rely more heavily on searching out evidence of human damage from additives. Obviously, this has merit as long as the government does not depend entirely on such evidence, and wait until harm shows up—which might not be for generations—before taking action. As Dr. Lederberg said, "It is not likely that we will and certainly we do not wish to learn much about the genetic hazards from the observation of catastrophes in human populations."

If we were just putting our first few additives on the market, we might be able to see a cause–effect association with harm in humans. But at this stage, with thousands of additives in use, it's impossible. That leaves, as an alternative, animal studies.

Controlled experiments in animals, the underpinnings of all toxicology, are at present our most reliable indicator of potential harm from food additives. We count on certain animals—which we assume have much in common with our own biological systems—to warn us of harm. Scientists feed the animals chemicals, inject them, paint them on, implant them, give them by stomach tube, and then observe what happens. The assumption is that what destroys living tissue in one species is likely to destroy it in another. Even so, these studies give us limited protection at best.

The Logic of Animal Studies

The first test done on laboratory animals, such as mice, rats, hamsters, rabbits, is usually the acute-toxicity test, in which several doses are given—orally or by injection—to determine how much of a chemical it takes to kill half of the animals in a short period, usually from twenty-four hours to two weeks. Whatever this dosage, it then is called the LD-50 (lethal dose for 50 percent). It tells how much of the chemical we can eat without quickly dying and may also give clues as to how the chemical reacts metabolically in the body and to its potential for serious damage. Next is the subacute study, usually lasting ninety days but sometimes six months; the purpose is to further define how much of a chemical animals can tolerate after repeated exposures for longer periods.

Most important, by far, are the chronic feeding studies, which last over the *entire* lifetime of the animal. Obviously, the most reliable study of additives is one that most closely matches the actual use by humans, by mouth and for a lifetime. In such tests the chemicals are administered orally to animals in various doses, in the diet or by stomach tube, for either way the chemical reaches the stomach and is subjected to digestive functions which may alter its biological activity, rendering it more or less harmful. Other methods, which may be appropriate for drugs, for example, injecting chemicals subcutaneously, intravenously or intramuscularly, may give skewed results because the chemical is quickly absorbed into the system without going through the digestive process.

The chronic feeding studies are sometimes called "lifetime feeding studies" or "two-year feeding studies," because that's the life expectancy of rats, the most often used animals, though mice, hamsters, rabbits, even dogs, pigs, monkeys may also be used. All animals—those that die in the interim or are sacrificed at the end of a predetermined period—must then be thoroughly examined, not only by the experimenter for visible harm but also by a specially trained pathologist who dissects the animals and examines all the vital organs under a microscope for any sign of damage. If the pathologist's work is sloppy, or if the animals are not examined thoroughly, as sometimes happens, crucial damage will go undetected.

The significance of lifetime studies cannot be over-emphasized, though few of the food additives we eat today have ever been subjected to lifetime feeding studies! As we've discovered time and again, the more serious hazards just don't show up in the short-term acute studies. In fact, results of acute studies can be misleading, often showing exceedingly low toxicity, and then the industry will go around for years saying how nonpoisonous the chemical in question is. (Cyclamate, for example, was touted by the industry as acutely nonpoisonous.) And then we discover later that over the long haul it's deadly.

Thiourea was a proposed preservative for citrus fruit and dried and frozen fruits. On the basis of acute animal studies the substance was declared relatively nontoxic; a few years later FDA scientists found it caused liver cancer in rats subjected to lifetime feeding studies, though the rats were quite resistant to acute poisoning. Similarly, ethylene glycol, diethylene glycol and polyoxyethylene monostereate showed no effects after ninety days' testing, but produced bladder stones when administered for a lifetime. Tetraethyl thiuram disulfide showed evidence of brain damage in a chronic study, but no such effects in a ninety-day study. Mercury compounds produced cumulative effects in the kidneys, and selenium in the liver, though neither showed such marked changes in short-term experiments.[7]

Birth defects and fetal deaths can often be uncovered in a short time by feeding mammals, usually rats, doses of a chemical during critical periods of pregnancy and observing the incidence of loss of embryos or of abnormalities in offspring. More definitive, however, and more pertinent to intake of food additives, are multigenerational studies in which the chemical is given through three or more generations of animals. Scientists then watch subsequent generations not only for birth defects, but also for fertility, infant mortality, viability of offspring, life spans and other effects which are applicable to humans but which would be overlooked in a less extensive experiment.

Another indicator of birth defects and fetal deaths is the chick-embryo study in which chemicals at various doses are injected into the egg sac or yolk of fertilized eggs. Since 1963, conducting chick-embryo studies at FDA has been primarily my responsibility. The test is relatively inexpensive, gives a large sampling (thousands of eggs can be used to test a single chemical), and is remarkable in that it can quickly—within twenty-one days, the hatching period for chickens—sound the alarm about potential causes of birth defects.

For example, testing of thalidomide in rats did not show damage to offspring. In fact, after birth defects were confirmed in humans, scientists could produce similar birth defects from the drug in only a certain strain of rabbits and monkeys—but not in rats, hamsters, mice, the usual test animals. Yet chicken eggs injected with thalidomide produced baby chicks with the same-type flippers found in human infants. This test could have warned scientists in twenty-one days of what it took epidemiologists five years to discover.

Despite this, FDA officials refer to the chick-embryo technique as a "screening process," and industry reviles it as "too sensitive" (which in industry jargon means it detects danger too readily to suit economic interests). They also criticize it because it is a "closed eggshell" system, in which the embryo is not attached to the mother as in humans, via the placenta. This is based largely on the old-fashioned assumption that the human placenta acts as a filter to keep chemicals out and that therefore any study on an animal without a placenta is not pertinent to humans. We now know that the placenta is not such a barrier; it lets all kinds of chemicals through to fetuses; in many cases a chemical concentrates more in the embryo than in the mother. That's why babies born of heroin addicts can emerge full-blown addicts themselves and immediately experience withdrawal symptoms.

Certainly chick-embryo studies, like other animal studies, are far from foolproof and should be considered in conjunction with tests on mammals. But, remembering the case of thalidomide, they should *never* be ignored, as they often are.

There are three methods for detecting mutations. In one, animals are fed large doses of chemicals, then examined for damaged chromosomes, the units of the cell which carry the genetic message. In another, animal cells are cultured in test tubes and treated with a chemical, then examined for broken or otherwise damaged chromosomes. The third is the "dominant lethal" method, in which male mice injected with large doses of a chemical are mated with untreated females; if there is a smaller number of live fetuses from the union than expected, it is assumed there must have been mutations in the sperm cells causing the reduction.

It's important to stress that although these are the main types of techniques generally accepted for testing food additives, they are not all being widely used, as we shall see later.

Which Man—You or Me?

Now, what the animal test results mean to humans is a complex and valid question, debated at any toxicological meeting. How do we know that animals metabolize, or burn up, the chemical the same way humans do? Perhaps their metabolic pathway is different and therefore the chemical is handled in an entirely different way in a rodent's body than in a human's. Do the animals in question store the chemical, excrete it, absorb it the same way as humans? Is injecting or implanting chemicals really valid when that is not the way the chemical gets into humans? Are the doses realistic when humans eat nowhere near that much of a chemical in a lifetime? Are the numbers of animals used sufficient? What about species' resistance and individual susceptibility?

The question boils down to: how similar is the testing of laboratory animals and the human situation? And what therefore can we conclude from such tests? At this stage of scientific advancement, the only honest answer is that we just don't know, and that it will probably be a long time before we do. The tests are, as the industry claims, crude, not always reliable and certainly not foolproof. But the alternative is: if animals are not studied as clues to danger, then what? Perhaps there will be another way in the future as our toxicological sophistication grows, but for the moment we are stuck with animal studies and we should be glad to have them.

It should also be pointed out that those who object most vehemently to the application of animal studies to humans do so often when results reveal economically harmful information. To quote James Turner, in testimony before a Congressional committee: "Most studies that are conducted on animals tend to show that the chemicals are safe and then are embraced by the producers of the chemicals and the FDA. However, when a study begins to suggest there are serious problems, immediate arguments are raised about the relevance of animal testing to human use."

All this is not to imply that we know nothing about the correlation of animal studies to humans. On the contrary, the literature is full of examples in which chemicals known to harm humans also harmed animals and vice versa. Such chemicals as soot, coal tar, creosote, mineral oils, tobacco were first discovered to cause cancer in humans and then tested out and confirmed in animals. In fact, no chemical known to cause cancer in humans, with the exception of one form of arsenic, has *failed* to cause cancer in animals. The probability that a chemical that harms animals will also harm humans is great. J. A. Miller, a cancer specialist, says that it is "almost incredible that our species could be exempt from such causes of cancer when other species are susceptible."[8]

On the other hand, certain animals used in laboratory experiments may be resistant to damage from certain chemicals that affect humans—for example, the mice, rats and hamsters that were used to try to confirm the assaults of thalidomide. This does not invalidate the extrapolation of animal data to humans; rather, it points up that such tests are not foolproof when they can't pick up a teratogen as big as thalidomide. This reinforces the argument that *any* indication of harm in any animal—even the tiniest clues—may be overwhelmingly significant. Dr. Lederberg takes a strict view: "There is no basis for the illusion that a compound that can cause harm in other living cells will not affect human cells."

In fact, there is considerable evidence that instead of overstating the potential risk, animal studies *understate* it.

First, humans are generally recognized to be more sensitive

to chemical harm than animals. According to information compiled by A. J. Lehman, former director of the Division of Pharmacology of the FDA's Bureau of Biological and Physical Sciences, humans are twice as sensitive to drug toxicity as horses and swine, three times as sensitive as cows, sheep and goats, six times as sensitive as dogs, and ten times as sensitive as cats as well as rats, the most common test mammals. Even far greater variations can exist in resistance, depending on the chemical and the genetic strain of animal used (one strain of rat may be more susceptible or resistant to a chemical than another strain). After-the-fact testing found that women were sixty times more sensitive to thalidomide than mice, one hundred times more sensitive than rats, two hundred times more sensitive than dogs, and seven hundred times more sensitive than hamsters.[9]

Second, animals used for testing are not representative of the frailties of the human population, which render us more vulnerable to chemical ravages. Test animals are the cream of the crop—superbreeds. They are specially bred, healthy (you discard sick animals); they're on a diet full of nutrients purified as much as possible to eliminate contaminants; they're not taking drugs or breathing smog; they're not subject to the environmental stresses humans are. They live in an air-conditioned, constant-temperature controlled environment called Utopia. This is obviously not the case with the human population, where you have the weak and the puny, the sick and the undernourished, the elderly and the stress-ridden, all eating foods and breathing air loaded with other chemicals. All of these variables can influence chemical effects; studies show that animals as well as humans subjected to stress, temperature changes and nutrient deficiencies are more susceptible to chemical assaults. The same has been found in humans with respect to cancer. Why do some smokers get cancer while others do not?

When asked how animal studies apply to man, Dr. John H. Weisburger of the National Cancer Institute replies, "Which man? You or me?"—so great is individual variance in susceptibility. Dr. Saffiotti has stated: "It has been found that . . . changes in age, sex, hormonal status, diet and nutritional status, genetic factors, individual variations in the metabolic handling of chemicals, or the combined effects of different chemicals all can alter the response of an individual to a chemical carcinogen; differences in response as high as 100-fold or 1,000-fold can be obtained by changing only one factor at a time."[10]

Obviously, test animals must be homogeneous—though they too have individual susceptibilities—and must be free from extraneous influences, so that we can be sure the tested chemical is indeed the factor causing the harm. But, precisely because we use only the elite and pampered of a species, it is a rare event when their resistance is overcome by a chemical.

Third, because of the small number of animals used in testing, it is far easier to miss a harmful chemical than to spot it. Imagine the problem of housing and the cost of using 200 million animals for a single experiment, though theoretically this would provide the most reliable results possible—on a one-to-one animal–people basis. However, most animal experiments use only twenty-five to a hundred rats or mice; and if monkeys or other primates are used the figure is much lower —perhaps seven to ten or so because of the expense. Now, these small numbers of animals, let's say fifty mice for example, are supposed to represent the total population of over 200 million Americans—each mouse standing for four million persons. Considering the vast differences in individual susceptibility of humans, it is impossible that one mouse could contain all the vagaries in his body that might plague some four million diverse persons in the country.

To try to compensate for the small numbers and the short life span of mice, scientists give doses a hundred or even a thousand times higher than those given to humans. These massive doses are often ridiculed by the food industry and inexcusably by FDA officials themselves as irrelevant to humans. When cyclamate was banned, some persons contended you would have to drink bathtubs of soft drinks every day to ingest cyclamate comparable to the great amounts that had caused cancer in experimental rats. That is pure nonsense. The well-accepted toxicological theory is that to compensate for the small numbers of animals used in experiments, massive doses are necessary to elicit all of the possible adverse effects that might occur throughout the animal's lifetime.

But even using massive doses in small numbers of animals, it is easy to miss chemically induced damage that could have enormous consequences to the population. Suppose a chemical is a carcinogen of such potency that it is inciting cancer in one percent of the population. Since one percent of fifty experimental rats is half a rat, such cancer would be totally overlooked in an ordinary animal experiment—and yet on the human scale the results could be disastrous. One percent of our population is two million persons!

The situation becomes more precarious when you're trying to spot harmful chemicals that may affect only fractions of percentages of animals and people. A chemical that provoked high blood pressure or liver damage in only one twentieth of one percent of the population would be virtually impossible to detect through routine animal studies. In reality, 400,000 persons in this country would be adversely affected. The same is true with weak carcinogens. Dr. Saffiotti says that if you pick up a carcinogen from present animal testing you can be sure it is a strong one; the weak ones just don't show up. Working under such limited circumstances it then is pure chance when we spot a chemical that is not superextreme in its harmful potency.

Because of the nature of cancer-causing agents, no "safe" level can be set for them. But with chemicals displaying ordinary chronic toxicity, toxicologists try to interpret "safe" dosages for humans—tolerances—based on the amount of chemical it took to harm the animals. Theoretically, in valid experiments the more of a chemical you give, the greater the harm. Ten micrograms of a chemical is supposed to produce roughly ten times as much harm as one microgram. That is why in an experiment you administer several dosage levels, graduated down the scale, to a group of chosen animals and their controls. Then you can tell precisely when you get to the point where no statistically significant damage shows up. (Controls—untreated animals—which are included in every experiment, may develop tumors and other diseases as they age, and these have to be considered and appraised statistically to determine any difference.) The point at which no damage occurs is called the "no effect" level or "safe dosage" for animals.

To set tolerances, toxicologists, however, do not apply the "safe" animal dosage directly to humans. If a rat weighing one pound could tolerate one ounce per day of chemical without exhibiting harm, it does *not* mean that a 120-pound woman could safely eat 120 ounces, or about seven and a half pounds, of the stuff daily. Here's where a lot of novices get into trouble—as well as knowledgeable persons trying to mislead the public. When they simply shift the figures to humans, the matter looks absurd. But it isn't scientifically.

Knowing that humans are more sensitive to chemicals than animals and that there is a wide gulf in individual susceptibility, scientists in extrapolating test results build in a "safety margin." They multiply the safe animal dose by ten to accommodate the species sensitivity difference; then they up that figure ten times again to make up for individual susceptibility. This adds up to a hundredfold safety margin, a figure firmly accepted in toxicological circles, though admittedly it is based on the crudest guesswork and has little real scientific basis. In truth, from what we know, the margin could well be greater, considering all the unknowns. No such extrapolation of one hundred, for example, would have helped the women who were two hundred times more sensitive to thalidomide.

But as a general rule—approved by the World Health Organization and written into FDA regulations—the hundredfold margin is used, usually expressed in parts per million of the chemical per kilogram of body weight. If an animal shows no reaction to a chemical at one hundred parts per million in the diet, it is assumed the chemical is safe for humans at only one part per million in the diet—or one hundred times less.

The World Series Approach to Science

We now can see that the more species of animals you test, the greater the chance of picking up hazard. For validity, the tests must be done on at least two species, including males and females. But what if one test shows up positive and the other negative? Or one test is positive and three are negative? Or four are positive and two negative? There are all kinds of ways to juggle scientific results, depending on the species you use (perhaps a certain strain is extraordinarily resistant to breast tumors), the stage at which you give the chemical (in one positive study of Red 2 food dye the chemical was given from conception, in negative studies the chemical was administered only after the sixth day of pregnancy), the time you look for damage (if you know brain lesions generally show up only after six hours in newborn but you look for them sooner, you won't find them), and other factors. Theoretically, however, experiments should be able to be duplicated; ones conducted precisely the same way should show very similar results.

It is essential to understand, though, that studies with negative findings do not cancel out the significance of positive studies, though that is the approach of many industry representatives. The FDA comes up with a study showing that mice get thyroid disturbances from eating chemical X, and the industry comes up with one showing that three mice in Kenosha ate the same thing and nothing happened. Then they assert that the first test showing danger is inconclusive or is insufficient basis for making a regulatory decision, or that the studies are "conflicting" or "contradictory" and thus worthless entirely. If more positive data show up, they find more negative data. And so it goes—"Let's go for two out of three, four out of seven, five out of nine." Then there is not even assurance that, arithmetically, numbers will win out. Negative studies sometimes triumph despite overwhelming positive evidence. Dr. John Olney, professor of psychiatry at Washington University Medical School, tells how in the FDA's eyes four negative studies on MSG (monosodium glutamate) "washed out" thirteen positive studies showing a wide range of dangers. MSG is still on the government's safe list.

This World Series approach to science and regulatory decisions is totally unjustified. Considering the crudeness of our testing, a positive result must be taken extremely seriously. FDA scientists consistently state in principle that when in doubt we must rely for maximum safety on the animal *most* sensitive and assume that man is at least equally sensitive. Thus, any animals that exhibit damage—under well-conducted experiments—indicate that something is wrong, negative evidence notwithstanding. Negative studies simply cannot be given equal or more equal weight in the presence of contrary positive evidence. Particularly in cancer testing, Dr. Saffiotti points out, weak cancer-causing agents would show up negative because current tests fail to detect cancer incidence below 5 to 10 percent. "Accordingly," he stresses, "we must pay great attention to the warning signal represented by a positive animal test."

The overriding question is, considering the crudeness of our testing methods, the near-impossibility of discovering harm from human experience, and the frivolous use of many food additives: Why take the risk? What harm can there be in a conservative approach to protection, as Congress intended? Government officials often rationalize their laxness by saying that precipitous action against food additives would be rash on the basis of "so little evidence." Considering everything, it is not caution that is rash but the government's failure to stringently enforce the law and to protect us against possible future catastrophe.

FDA Decision-Making

The logical question as a result of all this is: Why doesn't the FDA clean up the food supply? Why must we be subjected to so many untested or unsafe chemicals in our diet, many of which if taken in pill form could be obtained only by prescription? The simple answer is that the FDA is not performing its public function. Legally, it has the authority to end food pollution, or at least reduce it and the risks drastically, but it doesn't. Often the FDA seems more preoccupied with perpetuating the use of food additives than with getting rid of them—even though they are found unsafe. Some observers of the Washington scene laughingly say "FDA" stands for "foot-dragging artists."

The law is clear. The government does not have to prove additives unsafe. The burden is on industry to prove additives safe. If they can't do that, the additives should go out the window. But it doesn't work that way, because the prosecutor, judge and jury—that is, the FDA—sometimes shifts sides, assumes a defensive posture and lets the industry bulldoze or coax it into taking a stand detrimental to the public interest.

The main authority of the FDA over food additives stems from the federal Food, Drug and Cosmetic Act of 1938 and the Food Additive Amendments of 1958.[1] Under the 1938 law, the government had to prove a substance unsafe before it could force its removal from food. That was changed under the 1958 amendments, spearheaded by James D. Delaney, Congressman from New York, to bring the preclearance of additives in line with similar preclearance requirements for drugs. Now the law reads that manufacturers must show "proof of a reasonable certainty that no harm will result from the proposed use of an additive." Thus, the burden of proof for establishing safety was shifted from the government to the users who reaped the economic benefits. Since that time the creators of new additives have had to submit proof of safety, as outlined by the FDA, before the additive could be placed under "a food additive order." Using an additive without such approval could make a food "adulterated" and subject to government seizure.

However, there were loopholes. By the time the 1958 law was passed, there were already a lot of additives in use. Wiley's optimistic report of 1906 that food manufacturers had abandoned the use of "drugs in food" was short-lived, and additives had proliferated. Also, there were such common "additives" as sugar, salt, spices, used for centuries, which congressmen felt had to be allowed. So the country was stuck with a whole big group in common use which, after making some perfunctory inquiries around the medical community, FDA officials declared "generally recognized as safe," or GRAS (pronounced "grass"). These listed items, then—at the time, 182—as allowed by law, were exempt from food additive orders.

The GRAS items were not supposed to have detrimental information (to man or animal) against them in the medical literature, although, as it was revealed later, few of the medical authorities consulted by FDA even made a search to find out. Some opinions against certain proposed GRAS items were summarily scratched out by FDA officials as "unscientific" or "hearsay"—for example, one physician's opinion that cyclamate was unsafe.[2] Since few of these additives had ever been tested, the scientific journals were blank about their safety, and what studies had been done were probably locked away (especially if they showed harm) in industry files or buried somewhere at FDA, unpublished. The fact these items had been used—sometimes for centuries—was no reliable indicator of their safety.

So, your government's decision to make an item GRAS was hardly a scientific judgment you would want to stake your life on, and, not surprisingly, many of the GRAS items, when finally rigorously tested, have turned out to be dangerous, such as cyclamate, safrole, saccharin, brominated vegetable oil. And who knows how many others may fail, if and when we ever get around to testing them?

Furthermore, through the years the GRAS list grew as manufacturers came up with more candidates; sometimes the FDA knew about the items and added them officially to the list, sometimes not. So now nobody knows for *sure* how many GRAS items are in use, though there is a list of about six hundred published in the *Federal Register*. James S. Turner noted in *The Chemical Feast* that the figure is so vague that during the same period FDA officials reported 718, 575 and 680 items on the list. Since the definition of GRAS was not well controlled by FDA, the industry interpreted the law to mean that they alone could decide whether an additive was generally recognized by the scientific community as safe, use it and never feel compelled to notify the FDA of it. Incredibly, the FDA went along with this interpretation. Winton B. Rankin, then Deputy Commissioner of the FDA, said in October 1969, "The manufacturer is entitled to reach his own conclusions, based on his scientific evidence, that a substance is, in fact, generally recognized as safe. And he is not required to come to us then and get the material added to the list." Thus, the government let control of the GRAS list slip from its hands into industry's, and only now, under consumer pressure, is the FDA trying to regain control over it.

Another quirky problem is "prior sanctions." Under the 1958 law, another way industry could get permission to use additives was to write to the FDA and tell it in effect that there was nothing wrong with an additive they had been using and ask the FDA to send them a letter confirming that it was okay to go ahead using it. Nobody knows how many of these letters through the years went out from FDA, because they weren't filed in any central place and now can't be found. But manufacturers, when challenged, keep popping up with letters from FDA—and according to FDA's general counsel, Peter B. Hutt, these prior sanctions can't be revoked, because they too, like GRAS items, are exempt from food additive orders. In 1970 the FDA announced it was revoking all prior sanction *letters*, but *not* the prior sanctions themselves, and what that legally confusing statement means is anybody's guess. Except the end result is that manufacturers can still legally use additives because once upon a time they received prior-sanction approval from FDA.

Nitrite, for example, is a prior-sanction additive, as is SAPP (sodium acid pyrophosphate), recently added for the curing of hot dogs and other cooked sausages. In March 1972 the U. S. Department of Agriculture petitioned to allow SAPP for such use, and the FDA ruled that the additive would, as other new food additives do, require the filing of a food additive petition, showing it had been tested and proved safe for that use. But Griffith Labs, the producers, then informed FDA that they had received a letter of prior sanction in 1962 and had, in fact, been selling SAPP for use in cooked sausages since 1952, six years prior to the Food Additive Amendments. Thus, SAPP is now in hot dogs as well as other sausages, despite its lack of safety testing.

Generally the law gives FDA much discretion in determining what is "safe." However, an additive is not either always in or always out. GRAS items, however, are considered so safe they are unrestricted and can be used in any amount in any food. For additives under "food additive orders" the FDA can, if it thinks it necessary, set tolerances (so many parts per million in food) and prescribe specified uses. According to a House Committee on Interstate and Foreign Commerce, reporting out the bill:

> In determining the "safety" of an additive, scientists must take into consideration the cumulative effect of such additive in the diet of man or animals over their respective life spans together with any chemically or pharmacologically related substances in such diet. Thus, the safety of a given additive involves informed judgments based on educated estimates by scientists and experts of the anticipated ingestion of an additive by man and animals under likely patterns of use.

Cancer-causing additives are the one striking exception. The FDA was given no discretion, no room for shilly-shallying, on the question of cancer vis-à-vis food additives. Worldwide authorities on carcinogenesis (the study of cancer) time and again stress that there is no such thing as safely eating a little bit of a chemical known to cause a cancer in man or animals. A safe tolerance simply cannot be established on the basis of chronic-toxicity studies in animals, as for some diseases (though the FDA in its inferior wisdom sometimes ignores and disputes this).

Fortunately, this was codified by Congress in the 1958 Food Additive Amendments. A clause was inserted, sponsored by and named after Congressman Delaney, which prohibited the use of any food additive shown to cause cancer when fed to either animals or humans. Thus if the FDA discovers that a food chemical causes cancer, as determined by legitimately conducted studies, it *must* ban its use in food; the government cannot make any discretionary excuses.

The intent of the 1958 law is clear, as outlined by the House committee report: "to protect the public health . . . to prohibit the use in food of additives which have not been adequately tested to establish their safety . . ." The only question in regulating food additives is: Is it safe or unsafe? Or is it safe used in X quantities? That is all. The FDA does not have the authority to make judgments about whether an additive is desirable or necessary (though it can refuse clearance to additives deemed deceptive) or whether, if it is removed, harm will result to some vested interests in the country.

In short, the FDA does not have the power to allow or disallow the use of additives after trading off risk against benefit. Government authorities cannot say, "Well, this is of value to consumers—or industry—so we can leave it on the market, even though it is not safe, because, in our opinion, the benefits to society outweigh the risks." Legislators clearly displayed their opposition to any such determination by saying that the Secretary of Health, Education and Welfare (legal enforcer of the law) should make no judgment "of whether such effect [of an additive] results in any added 'value' to the consumer of such food or enhances the marketability from a merchandising point of view."[3] Although FDA administrators don't seem to have total recall on much of the 1958 law, this is a provision about which they have a persistent memory block.

Somehow—through lobbying efforts—color additives were omitted from the 1958 law, but they were brought under similar regulations through the Color Additive Amendments of 1960. And in 1968, federal legislation was updated again, bringing under FDA control medicated feeds fed to food-producing animals, which might leave residues in the meat. But on additives put directly into meat and poultry—for example nitrite—the FDA maintains only advisory capacity. These additives were left under the jurisdiction of the U. S. Department of Agriculture.

Under the law the FDA is also charged with protecting us from "indirect" additives, such as packaging chemicals that may migrate into food and accidental environmental contaminants, such as industrial chemicals (PCBs) and natural molds (aflatoxin). It is the FDA's responsibility to decide how much of such chemicals may be tolerated in food, to monitor the residues, to condemn the food if necessary and to prosecute the violators. The Delaney anticancer clause has not been interpreted as applying to such situations.

So though there are innumerable intricacies to the law, the essential ways of keeping unproved additives out of food are (1) adequate testing of new additives and (2) the swift removal from the legal list of additives which because of new evidence can no longer be regarded as safe. Animal testing as a proper method of detection of danger is also recognized in the law. What the government does with this authority, however, is another matter. You might say the FDA—and the USDA—operate on two premises: "You can't find what you're not looking for," and "It takes a mountain of data to raise suspicion about an additive's safety."

Lack of Testing

We know that many additives in common use have never been adequately tested for safety, except perhaps for acute toxicity, because they were in existence long before the Food Additive Amendments of 1958 were thought of—and some, especially those on the GRAS list, are now being tested for some aspects of safety. But even additives that have come on the market since 1958 and are now coming on the market may not be adequately screened. The FDA does not routinely require tests for long-term reproductive harm or mutations (though it does now require some testing for birth defects), nor, in a few cases, even for long-term chronic toxicity. FDA toxicologists make a judgment about whether an additive needs to be tested for cancer incitement on the basis of whether it is "toxicologically insignificant"—that is, is it to be used in small amounts in food not consumed on a large scale? If so a toxicologist may decide that a lifetime feeding study is not worthwhile because the chemical is "toxicologically insignificant"—is incapable of causing harm, and demands only a subchronic toxicity test which will not pick up cancer or a number of other diseases.

The concept "toxicologically insignificant" is accepted by a few scientists, but not by cancer experts at the National Cancer Institute. They say there's no such thing when it comes to cancer and chemicals. It is illegal to use additives which are known to cause cancer when fed to animals. But how can you prohibit a food additive on that basis if you don't even *ask* if it can cause cancer? Certainly you can't strictly enforce the Delaney clause.

The official excuse for not requiring mutagenicity tests is that they are not reliable, though this is disputed by a number of experts. In fact, four expert advisory committees have recommended mandatory mutagenicity testing for food additives, using present test methods.[4]

Certainly industry is entitled to know specifically what tests and test methods are required of it by the FDA. But indisputably, through scientific advances, some additives we thought safe on the basis of today's tests may turn out harmful when tested by more sophisticated methods in the future. Conversely, it's entirely possible that some additives now on the blacklist might be shown to be safe. As industry is rightly fond of saying in trying to trip up critics, "Nothing can be proved totally safe for all time." But the point is, additives can be and should be judged only on the basis of *today's best* scientific knowledge. After all, the worst that could happen is that by overestimating an additive's toxicity we are temporarily deprived use of an additive, one we may not need or want anyway. On the other hand, if the tests spot true danger, we would be spared long-term damage.

It is bad enough to consume untested additives. But it is more distressing to be asked to eat additives that are suspect or exhibit danger. Can something still be "generally recognized as safe" when there are a number of unanswered questions about its safety or scientific evidence against it? Consumer advocates argue "No," their definition of "safe" being "not apt or able to cause danger or harm." Again the intent of the law becomes important, as further delineated by the House committee report on the bill: to make manufacturers "pretest any potentially unsafe substances which are to be added to food." Preclearance or pretesting, then, is the legal intent.

In a nutshell, through our elected representatives we have declared that we do not want to accept *any* risks from additives. We want them to be prejudged safe—without hazard. We did not say we wanted to take certain risks as determined by the FDA. If indeed we as a society want to accept some risks, then they should be spelled out (is exposing one out of, say, 100,000 people to cancer, permissible?) and the law changed. In the meantime if the FDA were legally honoring our wishes as expressed by law, when hazards show up in additives previously thought safe it would not dawdle but would take immediate action to remove them from the food supply. By permitting industry to continue use of the additive, pending tests to prove it safe, the FDA is in clear violation of the pretesting intent.

Sometimes it is a case of simple delay—as with food colors. Nearly every high-school chemistry student knows that the term "coal-tar dye" is synonymous with hazard. And most of the two thousand tons of food colors we use each year are coal-tar derivatives. They were discovered in 1856 by William Perkin, then himself an eighteen-year-old schoolboy, who subsequently earned fame and fortune through his discovery. The dyes he synthesized (by mixing potassium chromate and aniline, a coal-tar material) were almost immediately snatched up by the fabric industry because they gave such brilliant,

lasting hues—purple, blue, green, red, yellow. It wasn't long
before the artificial dyes found their way into food, and soon
afterward cases of people being poisoned by the dyes began
showing up in the medical literature. The problem at the
time was mostly impurities, but as early as 1900 there was
worldwide suspicion that the coal-tar dyes might have toxic
properties of their own. Eighty were being used in this
country then, but, under the stringent supervision of H. W.
Wiley, only seven got on the FDA's approved list in 1907,
after they were tested on dogs and rabbits for acute toxicity.

As the demand for artificial color grew, new dyes were
listed, and, alas, some had to be delisted when it became apparent that they were unsafe: Green 1, Orange 1 and 2, Yellow 1, 2, 3, 4, and Red 4. Red 4 was subsequently given a
provisional listing, allowing it to be used to color maraschino
cherries. It is hardly a notable record of safety. As Dr. Jacobson has commented, "The history of approved dyes reads like
a guest register in a hotel for transients."

With the Color Additive Amendments of 1960, everyone thought the question of the safety of coal-tar dyes would
be answered once and for all. But they did not reckon with
the foot-dragging agility of the FDA. The law had, as one industry spokesman put it, a "permanent part" and a "temporary part." The "permanent part" called for testing of the
dyes to prove them safe before they could be added to food.
However, the "temporary part" allowed a "provisional listing"
for certain dyes, enabling manufacturers to continue use of
the dyes "pending the completion of the scientific investigations needed as a basis for making determinations" about
their safety. Originally, two and a half years were granted for
the testing to be completed; thus, all of the dyes should have
been evaluated—and approved or junked—by the end of 1962.
But under the law, the FDA was permitted also to grant "extensions" for the completion of the scientific tests. And this
it has done with frustrating regularity from that day to this.
Since the Color Additive Amendments were passed—at this
writing, thirteen years ago!—only two of the ten dyes in use
have qualified for a "permanent" listing on the basis of safety
tests. All of the others are still only "provisionally" listed without proof that they are safe.

There have been yearly extensions of the "provisional" listing for more than a decade, because, as one industry publication explained it, "difficulties have arisen in obtaining permanent listings for color additives." Understandably, because many simply won't qualify. In a large-scale FDA study of the dyes, Yellow 6 was found to cause blindness in dogs. Violet 1, used to stamp meat and put into some foods, is suspected, according to a Canadian study, of causing cancer. Here is what an FAO/WHO Expert Committee in 1969 said of Citrus Red 2, the color used to dye the skins of oranges: "Citrus Red 2 has been shown to have carcinogenic activity and the toxicological data available were inadequate to allow the determination of a safe limit; the Committee therefore recommends that it should not be used as a food color." The clincher came when the Russians reported in 1970 that Red 2, used under a "provisional" listing in this country, might cause cancer and reproductive harm.

In September 1971, under pressure, the FDA decided that it was time once again to put the coal-tar dyes under scrutiny, and in a "get tough" ultimatum it gave industry another two years to come up with safety data on the dyes. This time, however, the FDA could not—perhaps in good conscience and with everyone looking—renew the "provisional listing" for Red 2 in 1972, because of new evidence, both from the Russians and from new FDA studies, showing potential harm. So instead the FDA did nothing; that is, though failing to grant a provisional listing, it took no action to prohibit continued use of the dye by manufacturers. Attorney Anita Johnson of the Health Research Group calls this FDA silence "a blatant flouting of the law," claiming that the dye then is under no regulation whatsoever. The FDA argues that legally its failure to renew the Red 2 listing was meaningless (a strange admission), since the dyes automatically remain "provisionally listed" unless the FDA publishes an order revoking the provisional status.

For at least a century, these dyes have been under suspicion. It is now thirteen years since a law was passed demanding a show of their safety. During that time we have been eating food dyes at a rapidly increasing rate. In 1960, Americans consumed 2,159,000 pounds of dye. By 1972 the figure had almost doubled, to 4,000,000 pounds. Ironically, all food dyes are used only to make foods artificially more attractive, sometimes deceptively so, and to render them more devoid of natural flavor and color.

When Is GRAS not GRAS?

The FDA's largest problem is the GRAS list, now "under review." Under contract the FDA is having monographs compiled from the scientific literature on GRAS items and having some items screened for birth defects and mutations in animals. Incredibly, the FDA is *not* having GRAS items screened for cancer potential! The big question is: how will the FDA use information now flowing in on possible hazards? The FDA has not said that they will use the generated data to ban additives shown unsafe. Dr. Virgil Wodicka, director of the FDA's Bureau of Foods, has hinted that the test methods used on GRAS items may not be reliable enough to depend on. He said, "We have several contracts under way to test a selected number of the materials [additives] for teratogenicity and mutagenicity, as much to test the methods as to test the substances. The novelty of these fields is such that the methods are not generally accepted or agreed to and we need to make sure that the methods are okay before we draw any conclusions from the results." The FDA, it would appear, is not committed to heeding the results.

That many GRAS items may not be safe is generally recognized. In a 1969 internal FDA memo assessing the safety of the GRAS list, several experts in the FDA's Bureau of Science found many items "suspect" and others about which there were "substantial grounds for concern." Included were MSG ("neurotoxicity in newborn mice"); saccharin ("possible bladder cancer"); brominated vegetable oils ("growth inhibition and organ pathology in rats"); ammoniated glycyrrhizin ("possible inducer of adrenal-like difficulties"); potassium bisulfite ("powerful reductants which may impair nutrients or introduce toxicants"); hydrogen peroxide ("powerful oxidant which may impair nutrients and produce changes in foods resembling those due to ionizing radiation"); carob bean gum, ghatti gum, guar gum, sterculia gum, gum tragacanth

("FAO/WHO Expert Committee refused recently to set acceptable daily intake levels on basis of available evidence"). Forty-eight other suspect items were listed, including benzoic acid, sodium benzoate, BHA, and most of the flavorings and spices, which have not been tested.

The FDA has data, never published, showing that allyl heptalate, a synthetic pineapple flavor, may cause liver damage when ingested by both rats and dogs. The rat study was completed in 1960, the dog study in 1963. All six of the dogs on high doses of allyl heptalate died of hemorrhage and cirrhosis of the liver within seven months after the study began—an alarming mortality rate. Yet the data remain buried and allyl heptalate continues to be used.

Another flavoring, gamma valerolactone, commonly used in ice cream, candy and baked goods, "is suspicious of being cancerogenic," according to an FDA pathologist who in 1970 analyzed the results of an FDA study completed in 1967. Rats that were fed the chemical for two years developed twenty-six malignant tumors compared with two tumors among controls. Because of an erratic dose response, scientists are troubled over the validity of the study; nevertheless, as Morris A. Weinberger, director of the Division of Pathology, pointed out, it is impossible to make "a positive statement that carcinogenicity for gamma valerolactone has been ruled out."

In mammalian teratology studies being done by Dr. Bernard Oser's Food and Drug Research Laboratories in Maspeth, New York (not connected with the FDA), preliminary evidence showed sodium acid pyrophosphate (SAPP), carrageenan and sodium benzoate to cause some fetal harm. Dr. Oser is screening a number of GRAS items for teratogenesis, using mice, rats, hamsters and rabbits.

Aspirin too, although it is obviously not a food, is getting a teratology run-through by Dr. Oser. At the insistence of the FDA, he is using aspirin as a control substance to determine whether or not the animals are susceptible to birth defects. And the reason he is using aspirin, in the words of one FDA official, is that the drug is such a "potent and reliable teratogen." It can be depended on to produce birth defects and fetal deaths over and over, as Dr. Oser has confirmed. His experiments show that the offspring of animals given aspirin during pregnancy are born with serious abnormalities—such as spina bifida, in which the spine is incomplete (a human born with this defect usually does not survive long), and exencephaly, in which the brain is exposed. The doses are high—between 150 and 200 milligrams per kilogram of body weight daily—but some humans do take large doses of aspirin. Also, when the doses were lowered, fetal deaths went down but abnormalities went up.

There is evidence going back fifteen years that aspirin may cause birth defects, but this is not generally known to the public. Aspirin is considered one of our safest drugs; it is sold to anyone and everyone, and pregnant women innocently gulp it down by the handful. Yet the FDA is so sure aspirin is a cause of birth defects that it uses the drug as a yardstick against which to measure the teratogenic power of food additives, and then it doesn't tell anyone but just quietly files away the information as it keeps coming in.

What happens to a GRAS item that becomes suspect? Since it can no longer be "generally recognized as safe," it is no longer legally entitled to exempt status and should be banned until it can meet safety testing requirements demanded of a food additive under a "food additive order." However, faced with banning many GRAS items under the new "review," the FDA in July 1970 invented a new legal gimmick, the "interim food additive order," to skirt the spirit of the law and save such chemicals.[5]

When evidence of hazard accumulates against a GRAS item to the extent that it can no longer be considered safe, the FDA may indeed sweep it off the GRAS list, often done with fanfare to illustrate tough action. You may think that means it's removed from food, but this is not so. It may merely have been shifted to another legal status, the "interim" order, under which industry can continue its use pending a promise to produce new studies showing the item safe within a specified period—usually two or three years. It's much like the "provisional" listing for colors. The chemical hasn't been proved safe; yet industry can still use it.

Essentially, the interim regulation is a grace period giving industry time to come up with studies showing safety or with conflicting studies to confuse the issue, or time to develop a substitute or reformulate their products to eliminate the harmful additive. In the meantime, you and I are guinea pigs as much as the animals in industry's labs.

Such is the case with brominated vegetable oil (BVO), which spurred creation of the first interim order. BVO for years has been used by makers of both carbonated and non-carbonated fruit-flavored beverages to keep flavoring oils in suspension and give the drink "body." Scientists had been uneasy about BVO because of a lack of long-term studies and a suspicion that bromine from the BVO might be stored in human tissue. Expressing these concerns, an FAO/WHO Expert Committee on Food Additives in 1966 noted that it didn't have enough evidence to evaluate the toxicity of BVO. But by 1970 the same committee said new evidence "suggests that a human epidemiological problem could arise from the uses of BVO" and that it "should not be used as a food additive in the absence of evidence indicating safety."

That new evidence came from two sources. The British Industrial Biological Research Association (BIBRA) found that animals did indeed store bromine in their tissues—notably in the liver, the kidneys and the heart, but also in the spleen, the pancreas and the lungs. Around the same time scientists at the Canadian Food and Drug Directorate, testing BVO in the diet of rats for eighty days, found that the chemical caused liver, heart, kidney, spleen and thyroid damage. The study was published in 1969. In January 1970 the Canadian Health Minister reduced the level of BVO allowed in beverages to fifteen parts per million (formerly it had been three hundred parts per million). Take note that other countries were more cautious. Belgium gave up BVO in 1967; Sweden banned it in 1968, as did Great Britain in 1970—on the heels of the new studies at home and in Canada.

The Canadian study and action caused a flurry of activity at the FDA. And Dr. Charles Edwards, newly arrived at the FDA as Commissioner* after the departure of Commissioner Ley over the cyclamate incident, made a tough pronouncement that a use- now, test-later philosophy would not be

* Dr. Edwards is now (1973) HEW's Assistant Secretary for Health.

tolerated for food additives. In January 1970, in a dramatic move, he banished BVO from the GRAS list; industry, he said, would have six months to eliminate it or come up with evidence for setting a safe level. He said of BVO at a press conference, "The public health cannot be endangered for months or for years while we attempt to accumulate all the scientific data needed for an absolute determination of safety or danger. Therefore, we will sometimes make decisions to regulate out of commerce suspect products, on the basis of demonstrated doubt, and we will not regulate them back into society until science has allayed these doubts."

And then what happened? The lobbying machinery of the food, beverage and chemical industries went to work. It convinced the FDA that the British and Canadian studies were inconclusive and promised that American industry would do its own tests—if given time. So, in July 1970, on the very day of the six-month deadline when BVO was to expire or be scientifically cleared, the FDA quietly published notice in the *Federal Register* that BVO would be authorized at fifteen parts per million in beverages and would be regulated under an unprecedented interim food additive order. The industry was given three years—until 1973—to complete tests under the interim arrangement either exonerating BVO or setting permanent tolerances on its use—which writer Daniel Zwerdling compared with "commissioning ARCO to do a study of whether or not the Alaska pipeline should be built." The interim order is precisely what Dr. Edwards had six months previously railed against.

The interim order has now become an established procedure for dealing with suspect food chemicals (saccharin was the second item to be thrown into that status), and the FDA intends to use it routinely after the review of the GRAS list is completed—in 1974 at a cost of some $20 million to taxpayers. Two- and three-inch-thick monographs (containing toxicological data from a search of the literature as well as new results from the teratology and mutagenecity studies) on 593 additives are to be submitted for review to the Federation of American Societies for Experimental Biology. Then the FDA will review and decide, say officials, whether an additive should be classified as GRAS, banned or put under a food additive order or an interim food additive order, pending further two- or three-year tests.

It's not difficult to envision that this process will buy up time until around 1980. And there's also the question—if industry is allowed to do the follow-up safety studies on interim additives—whether we'll be much better off than before. For it has not been unheard of for labs beholden to industry for their livelihood to consistently come up with sponsor-pleasing results.

Bucking Decisions to the Academy

Sometimes the FDA does not even make the decisions about food additives. Instead, it evades responsibility by referring the matter to the National Academy of Sciences, which then appoints a committee to study the available information on an additive and come up with a recommendation. And so the decision passes from the public arena, where it belongs, to a private one, and finally trickles down to a select group of seven or eight men who operate in secret and are not accountable to the public.

Though some FDA officials say that they dislike consulting the Academy and that they sometimes get bad advice, it is a consistent practice whenever a tough decision with great impact on the food industry must be made. Dr. Wodicka has said: "I don't like the idea of fobbing off our job on someone else, and I don't think the Academy especially likes it either. But the issues are so controversial we are trying to take them out of the area of special interest. It's like a court of last resort, although I wish we didn't get pushed there so often." Another top official believes that "a good proportion of FDA's problems are traceable to the Academy." He terms some of their recommendations "miserable." Although the FDA is not obligated to, it usually follows NAS advice. It is the path of least resistance and the path that usually leads us right to the wolf's door.

The FDA often has better scientific expertise than the men who sit on NAS committees. The National Academy of Sciences, it is true, is an august body, made up of some 950 distinguished scientists elected to it because of their formida-

ble scientific achievements. Many of them are Nobel Prize
winners, such as Dr. Lederberg. But these scientists rarely sit
on the Academy committees that make the judgments. The
Academy has what it calls its "working arm," the National
Research Council, under which are a number of standing
committees, including the Food Protection Committee.
When the FDA has a sticky food problem, it asks the Food
Protection Committee to set up an ad-hoc subcommittee to
study all the data available (which is supplied by the FDA)
and come up with conclusions and recommendations for
action.

The men appointed to the committees make grave decisions affecting public health, yet they often hold views contrary to those of consumer advocates and of FDA and National Cancer Institute scientists, and have close connections
with industry, which would be considered outrageous in FDA
personnel. Any FDA scientist making an evaluation on public
health who took money on the side for doing research for the
companies he was required to regulate would be quickly fired
and probably taken to court on conflict-of-interest charges.
No such strings of delicacy tie the hands of scientists working
within the Academy confines. Many on the Academy advisory
subcommittees are or have been employees of industry, or
are professors or independent laboratory personnel deriving
much of their livelihood from research contracts with
industry.

The Food Protection Committee is both supported by and
beholden to industry. In 1970–71 the food chemical and
packaging industries contributed $68,000 to the committee's
general administrative expenses. The committee also has an
Industry Liaison Committee, whose head, Arthur T
Schramm, says it was set up by the Academy to give industry
a "pragmatic input."

The Food Protection Committee's conflict of interest in
making public decisions is clearly spelled out in one of it
pamphlets: "The Food Protection Committee was formed
as an advisory group, its primary purpose being to provide
critical evaluation of information concerning the relation t
the public health of technologies used in food protection for
the counsel and encouragement of the food industry and a

guidance for public agencies." (Italics added.) At the time the pamphlet was issued, in 1970, the committee had a Toxicology Committee, headed by Dr. Julius Coon, professor of pharmacology at Thomas Jefferson University in Philadelphia; ten members served on this subcommittee, and five of them were receiving industry funds either as direct employees—such as D. W. Fassett, Eastman Kodak Company—or as directors of laboratories depending on industry money for research.

Not surprisingly, the thinking of the Academy subcommittee members reflects their orientations, and their advice to the FDA is invariably antagonistic to the law and to consumer interests. The consulting committees have made it clear they oppose both the point of view and certain specifics of the Food Additive Laws. The Food Protection Committee has gone on record as favoring repeal of the Delaney clause, and Dr. Coon, often a chairman of the subcommittees, personally calls the clause "scientifically preposterous." In 1970 the Food Protection Committee issued a report setting guidelines for "toxicologically insignificant" amounts of cancer-causing chemicals in foods, in direct contradiction of the Delaney provision. An ad-hoc committee of leading scientists under the auspices of the National Cancer Institute was so appalled by the report that in a rare gesture of scientific infighting it condemned it as a total misunderstanding of the problems of chemical carcinogenesis.[6]

Generally, the Academy subcommittees are picked from a clique of scientists well known for sharing the industry point of view on food chemicals. (They are sometimes jokingly referred to as the Hertz Rent-a-Scientists.) They can be counted on at scientific meetings to pop up with the view that additives are good for you, necessary for progress, and harmless. Says Dr. Coon, who chaired the Academy ad-hoc subcommittees on cyclamate, saccharin and Red 2: "There is not a shred of evidence or even a basis of reasonable suspicion that any such damaging effects have ever been caused by the additives or pesticides in food consumed in North America. Certainly some defects have been observed in test animals after they have been fed exceedingly large amounts of some additives. But it is a long, frequently too long, step from the observation of the effects of such provocative and bizarre experiments to those of man's daily diet."

Further indication of the character of judgment expected from these committees can be gleaned from the attitude of Dr. William J. Darby, former head of the Department of Biochemistry and director of the Division of Nutrition at Vanderbilt University School of Medicine, who for eighteen years, until his recent retirement, was chairman of the National Academy's Food Protection Committee. During his tenure Rachel Carson's *Silent Spring*, an eloquent plea for a sensible examination of pesticides, was published. Miss Carson's best-selling book was attacked from a number of sources, including the U. S. Department of Agriculture and establishment scientists. One of the most widely circulated and virulent attacks in a scientific magazine was written by Dr. Darby in *Chemical and Engineering News*. He wrote "Her ignorance or bias on some of the considerations throws doubt on her competence to judge policy," and then went on as Fred Graham in his book *Since Silent Spring* recounts, to refute statements Rachel Carson never made. Dr. Darby is now president of the Nutrition Foundation, an industry supported group, and is still a fervent crusader against what he calls "irresponsible, biased scientists" who "alarm the public unnecessarily."

Those who don't bend so easily to the prevailing opinion of the Academy committees don't get on them or aren't invited back. A former executive director of the Food Protection Committee, Paul Johnson, who held that position for fifteen years until 1971, said he couldn't remember a time when an actual *elected* member of the Academy served on working committee judging food safety. Dr. Wilhelm Heuper, formerly of the National Cancer Institute, was appointed to the Food Protection Committee for one term, but was not reappointed, in his view because he took a tough stand against cancer-causing additives. James Crow, a geneticist at the University of Wisconsin, was called in by an Academy subcommittee as a special consultant on the mutagenic potential of cyclamate, but when he disagreed with the committee members his views were totally ignored, though he is a topflight cientist in the field, and no mention of his dissenting opinon was included in the Academy report to the FDA.

Consequently, the panels are rarely made up of scientists s competent as others available to the government, and they requently do not include an authority in the field under tudy. It's not surprising that the reports of these Academy ubcommittees often contain scientific gibberish, embarrassng both to the prestigious Academy and to the FDA. Their ecommendations invariably follow a predictable pattern of do nothing now; there's no imminent danger, and we need nore study"—which is a safe bet. In the meantime the public s left exposed to the hazard in question. The Academy subommittees have seen little or no danger in a long string of dditives, including cyclamate, saccharin, MSG and the food lye Red 2, all of which at the time had substantial scientific vidence against them.

The subcommittees have been strongly criticized by conumer advocates such as Ralph Nader. And some Academy nembers privately find the subcommittees' recommendaions, which come out under the imprint of the prestigious cademy, embarrassing scientifically. To try to correct this, he Academy's president, Dr. Philip Handler, in 1971 established a board of Academy members which reviews the subommittees' reports prior to release. However, the reviewers an only recommend, and have no authority to compel their esser colleagues to change interpretations or advice. In fact, he ad-hoc committees are treated gingerly by the Academy taff, who are reluctant to offend them. Said one staffer, These men are volunteers; they donate their time, and the nly thing they get out of it is traveling expenses and the restige of associating with the Academy."

Surely the impropriety of referring public decisions to such AS committees is evident. As Dr. Epstein has said, "The lose identification of the NAS-NRC Food Protection Comittee with industrial interests make it singularly inapproprite as a major source of 'independent' advice."[7]

One has only to consider the scandal of monosodium glumate (MSG) to see how right he is. Anybody who has ever ad Chinese Restaurant Syndrome, caused by an overenthusitic use of MSG in Oriental food, knows that this chemical can't be all good for you. Though researchers say the condition is transient, causes no damage and affects only certain persons, it can be frightening when the numbness creeps up the back of your neck and you begin to have headaches and a tightening of the chest, typical symptoms of CRS, as it's called.

Scientists once thought that MSG, the sodium salt of glutamic acid, which is a naturally occurring amino acid, was one of the most harmless chemical substances. It was isolated from seaweed in 1908, and for years the white chemical substance has been made and sold in the U.S. by bulk (mostly under the label Accent). It's now in thousands of processed foods to "enhance flavor" and is virtually impossible to avoid. It's in canned vegetables and frozen foods galore, in TV dinners, all kinds of convenience foods, in soups, soup mixes, salad dressings, mayonnaise, baked goods, even potato chips and crackers.

MSG was also in baby foods until 1969, when baby-food manufacturers removed it on the heels of consumer charges that it was needless, deceptive and dangerous to infant nervous systems. Harvard nutritionist Dr. Jean Mayer said MSG had no nutritional value and was put in baby foods to please mothers' taste buds and to conceal the overuse of starches and the underuse of meats. The very day after Dr. Mayer told a women's press club audience, "I would take the damn stuff out of baby food," an agitated industry surrendered and removed it. Perhaps if we had waited for action from the FDA, it would still be in baby food. For the FDA had the same evidence as Dr. Mayer and Ralph Nader and Jim Turner that MSG was suspect, but made no move to curtail its use. In fact, to this day (1973) the FDA considers MSG "generally recognized as safe," despite staggering evidence against it.

Dr. John W. Olney, a psychiatrist at Washington University School of Medicine, first brought to light in 1969 the fact that MSG caused irreversible brain damage in infant animals. His studies spurred more studies, and in September 1972, before a Senate subcommittee chaired by Senator Nelson, Dr. Olney viewed what had happened since his original revelation:

MSG-induced brain damage has been demonstrated in infant mice, rats, chicks, monkeys. The effect occurs following oral as well as subcutaneous administration and at doses of MSG which do not differ in order of magnitude from those used in foods. Obesity and neuroendocrine disturbances following treatment of infant rodents with MSG have been confirmed by neuroendocrine specialists. A recent report from Japan, which has been confirmed in my laboratory, documents lesions induced in fetal brain by administration of MSG to pregnant mice. Behavioral disturbances, including learning deficits, have been shown in adult rats treated with MSG in infancy. It has also been shown that MSG penetrates and is accumulated in the particular brain region known to be most vulnerable to damage from this food chemical. Several additional compounds have now been found which produce the MSG-type lesion; each is similar to MSG in molecular structure and in neuroexcitatory action; each increases the neurotoxicity of MSG if administered with it.[8]

He documented this impressive evidence by citing eighteen studies, and added: "That MSG intake can irreversibly damage the brain of infant animals is well accepted now by any serious student of the neurosciences. Yet MSG remains on the GRAS list today to be used freely in any foods, in any amounts and for any age group. How is this explained?"

Dr. Olney has an explanation: "an industry-arranged whitewash" carried out by a largely industry-oriented committee of the National Academy of Sciences, appointed to assess the hazards of MSG.

"I remember it well," Maurice Chevalier might have sung. For, after Dr. Olney published his findings in the May 1969 *Science*, there was a scramble at the FDA to justify the presence of MSG on the GRAS list. In September of that year Commissioner Ley testified before the Senate Select Committee on Nutrition and Health along with Dr. Olney, and presented evidence from four studies showing that MSG was safe. The only hitch was that the commissioner had been misinformed, as Jim Turner found out and exposed: two of the studies he cited were incomplete and two did not exist.

One of the studies touted was one I had done with chicks to determine whether MSG caused fetal defects. When it was cited to the senators as proof of safety, the study had barely begun and the results were far from conclusive. I had tested only 180 eggs—a very preliminary study. Particularly disturbing was the fact that the policy-makers attached great credence to this partially finished test which apparently revealed no damage, while refusing at the same time to countenance an impeccable 13,000-egg study on cyclamate which revealed severe skeletal defects. It was a striking instance of selectively sifting scientific evidence to ratify agency policy.

Confronted with a messy situation, the FDA contracted with the National Academy (whose services are not free—the taxpayers pay for its advice) to set up a committee to evaluate MSG in baby food. In July 1970 the committee reported that "the risk associated with using MSG in foods for infants is extremely small." The report stated that since the committee couldn't find any benefit to infants from MSG, it was recommending that the chemical be eliminated from infant foods, but it added:

> The committee found no evidence of hazard from the responsible use of MSG in foods for older children and adults, except for those who are individually sensitive to the substance. The flavor-enhancing property of MSG is considered to be beneficial to the general consumer in these age groups. [Note: the committee is here weighing a benefit–risk trade-off which exceeds its charge and which is clearly forbidden by law from entering into FDA judgments.] The committee, therefore, recommends that use of MSG be permitted in processed foods for these groups and that such foods be clearly labeled to indicate the presence of added MSG for the information of those who wish to avoid it. Sale of MSG in packages for institutional and home consumer use need not be curtailed.

Dr. Olney says he tried to pound home to the committee the very slim margin of safety in his experiments—only eight to one. A one-pound monkey showed brain damage after being given the equivalent of only two to eight jars of baby food containing usual amounts of MSG. The mice he tested were a sight. Those given MSG daily in the first ten days of life were obese and stunted at nine months old—apparently because MSG damaged the part of the brain that controls the appetite mechanism. The MSG-fed mice were lethargic; some females were sterile; their livers, uteri and ovaries were affected. Dr. Olney noted that the small percentage of brain cells destroyed in the infants was "evidence of a subtle process of brain damage in the developmental period which could easily go unrecognized were it to occur in the human infant under routine circumstances."

Yet the NAS committee gave more weight to two negative studies on MSG hurriedly done by two laboratories which derive most of their income from industry-contracted research.

In his 1972 Senate testimony, Dr. Olney—at the senators' request—gave his views on this negative research, pointing out flaws that the NAS committee had failed to examine. He also said that there were "many reputable neuropathologists in the United States who could have been called upon to judge the authenticity of evidence,"—but except for one those chosen for the committee had little or no qualifications in neuropathology. But most shocking to the senators was Olney's disclosure that five of the seven members of the NAS committee that had exonerated MSG in 1970 had industry connections—some of them directly with MSG manufacturers or with baby-food companies.

The chairman of the Academy's MSG committee was Lloyd J. Filer, Jr., a professor of pediatrics at the University of Iowa Medical School, who while serving as chairman was under contract to do research on MSG paid for by Gerber Products, Inc., a leading baby-food manufacturer, and the International Mineral and Chemical Corporation, the producer of Accent and of 80 percent of the MSG sold in this country. Another committee member trusted with an objective appraisal of MSG was George M. Owen, a researcher at Children's Hospital in Columbus, Ohio, who shortly before his appointment to the committee had also done work for Gerber. Two other members were toxicologists for chemical companies—John A. Zapp, Jr., with Du Pont, and Virgil B. Robinson with Dow Chemical Company. Rounding out the industry representatives on the committee was Lloyd W. Hazelton of Hazelton Labs in Virginia, which also has done testing of MSG for International Mineral and Chemical. Hazelton also appeared at the Senate hearings as an industry spokesman on behalf of the Grocery Manufacturers Association.[9]

All of this caused Senator Charles Percy, a member of the

Senate committee, to explode: "Here is the same person acting as judge and jury. Regardless of his objectivity and competence it appears there is a conflict of interest here. Aren't
there enough scientists in the United States so we can select
a panel that would be above reproach?" Percy compared the
apparent conflict of interest on the Academy panels with that
forbidden by the Bar Association and the Supreme Court
and suggested a Congressional investigation of the NAS committee procedures. The National Academy was offered a
chance to send a spokesman to the hearings to testify on its
behalf, but the invitation was declined. Dr. Handler did
however, subsequently submit a document to the Senate committee, defending the Academy's procedures. He admitted
that the Academy did not prohibit persons from serving on
committees because of apparent conflict of interest.[10]

Obviously, with twenty studies at last count showing MSG
to possess hazard and four (all industry-supported) showing
no hazard, this chemical can't be generally recognized as
safe. Nor is the problem solved with its removal from baby
foods. As Dr. Olney stresses, it is still present in adult foods
at significant levels—in fact, a recent Academy survey revealed
that more MSG is used in the United States than any other
GRAS item—and these foods are given to infants also. Many
babies eat right from the table today, as a trend toward solid
food at earlier ages increases. What is the good of taking MSG
out of baby foods, if babies still get it and parents aren't
warned about its presence? There is no warning on food cans
or packages. Furthermore, whether MSG may be hazardous
to older children and adults and whether it may cause reproductive damage are unsettled questions.

In any event, the point remains that the FDA's decision-
making function is compromised when it fobs off decisions
on any group—particularly one that is private, secret, and can't
be held publicly accountable.

4

Industry-Government Coalition

Despite its legal mandate to safeguard the public health, the FDA, like other governmental agencies, becomes more identified with the industry it is regulating than the consumer it's supposed to be protecting. It's the bureaucratic disease, and it comes from too much elbow-rubbing. Many government officials assume that the consumer and the food industry have identical interests. What's good for one is good for the other, to paraphrase the onetime head of General Motors.

There's no disputing that industry has the ear of our government guardians far more than do consumers. At the top the pressures are considerable. It is reported that when Robert Finch, then Secretary of Health, Education and Welfare, was compelled by law to ban cyclamate, he was immediately contacted by Donald Kendall, head of Pepsi Cola (which used the artificial sweetener in its wide-selling Diet-Pepsi), who was also a personal friend of President Nixon's. For others, though, it is a sharing of like interests with industry personnel or buying a ticket for the future. Many FDA decision-makers either come from industry or plan to go there later. It's been called the "revolving-door" policy and the "deferred bribe." Commissioner Edwards, prior to taking the position of head of the FDA, was with Booze, Allen and Hamilton, management consultants. When his deputy, James D. Grant, resigned, he became assistant to the chairman of the Board of CPC International, which owns Best Foods. The director of the Bureau of Foods, Dr. Virgil Wodicka, was formerly with Hunt-Wesson Foods. When there was a recent flap over the food color Red 2, one of the men sent by industry to discuss the situation was the former chief of the colors division of FDA. The interchange of personnel is so legion that some meetings with industry representatives—of which there are many—resemble a reunion more than a conference. In a study called "Who's Who at FDA," Dr. Jacobson notes that twenty-two out of fifty-two top FDA officials were former employees of regulated industries or trade associations.[1] It is not unusual at scientific meetings for an FDA person to publicly defend industry if it is attacked.

Even the lower echelons at Food and Drug are exposed to constant contact with industry but are very much insulated from consumers. Industry representatives—from companies from laboratories doing industry-sponsored research, from trade associations—are constantly showing up at the door, either with or without appointments. They can walk into any office, be given data and be treated like friendly colleagues. But let a consumer come in and ask for the same information, and he is often treated with suspicion. Industry representatives can walk in, but consumers must fight their way in. Recently, industry people who were concerned about some damaging new FDA data on Red 2 asked to come in and discuss it with FDA scientists. They did. But when members of Ralph Nader's Health Research Group asked for the same privilege they were initially refused. Only after vigorous protest were they too permitted to see the ongoing studies. What is common knowledge to industry is often kept secret from the public. Though this has lessened under new interpretations of the Freedom of Information Act, the FDA still gripes about taking time to dig out the information; yet its staff spends hour after hour, day after day talking with industry. I personally have attended dozens of meetings with industry, but I have never been to a single one with consumer representatives.

There are other ways too of hiding information to try to keep it out of the public eye, circumventing the Freedom of Information Act. A new propensity has developed among some FDA officials to hold telephone conferences with industry, in which formal proceedings need not be kept in detail. In case something goes wrong or concessions are made, there are no tracks across the bureaucratic carpet with which the agency can be indicted. On particularly sensitive issues, there are two memos of meetings: a detailed one for the commissioner and another, a short summary, to be circulated among participants and made available for public information.

How little the consumer is listened to was illustrated by a recent reaction of the Department of Agriculture when Griffith Labs asked approval for use of the chemical SAPP in hot dogs and other sausage meats. The USDA received 447 public comments, nearly 90 percent of them from consumers who opposed approval because they didn't want more unnecessary additives in food. Even some scientists in Agriculture argued against using SAPP for accelerated curing, saying that the new reduced cycle might fail to kill dangerous organisms. Nevertheless industry wanted it (SAPP also cuts manufacturing costs and increases profits), and industry got it. Said Dr. Kenneth McEnroe of the Agriculture Department, "Consumers told us they didn't like SAPP and didn't want it, but they didn't provide any data that was truly useful and industry did."[2]

It is not that government decision-makers are corrupt, but that their sense of public duty is constantly eroded by industry contacts and the consideration of short-term effects on industry instead of long-term effects on consumers. Attorneys working with Ralph Nader captured the spirit of the situation precisely when they wrote:

> The food industry, its trade associations and research foundations, is well financed and highly organized to pressure the FDA. Industry representatives visit, phone and write FDA officials about every decision that could affect their financial position, constantly badgering, cajoling, presenting facts, figures, etc., to persuade officials to their points of view. There is no comparable input from the consumer side. Industry pressures gradually erode even the most vigilant FDA officials. In delicate situations where risk is suspect and not proved, industry presence makes the difference. If the FDA acts on the side of caution and pro-

hibits an additive, industry can quantify their losses in terms of a dollars and cents economic loss. Human loss, on the other hand, is never a definite sum of money.[3]

A possible catastrophe from cancer or mutations to some unknown persons far in the future can't be traced to one official's mistake, whereas an immediate one affecting industry decidedly can. Confronted with threats to his advancement and little or no pressure or rewards from the consumer side, the policy-maker may give in. If he resists, he may actually be punished by his superiors or colleagues for being "overly conscientious," "rash" or "irresponsible" or "unethical" for "unduly alarming the public" or "making the agency look bad."

Particularly disturbing is that both industry and FDA play down the danger from food additives. It is common for spokesmen from both to lament the "crisis in confidence in the food supply," which consumer advocates and some scientists bring. Dr. William Darby, president of the industry supported Nutrition Foundation, has on several occasions criticized the "premature, nonobjective, biased statements" of scientists "who undermine the public confidence in the food supply." It is, he contends, "a betrayal of trust." FDA personnel at meetings have also echoed that view.

There is justifiable reason for a loss of confidence in the FDA and the food supply. What is the public to think after years of assurance that its food supply is the safest, upon learning that food is not free of filth—that there are only "filth guidelines" allowing certain percentages of insects, rodent droppings, mold and other debris in food; that fish contain unacceptable levels of mercury; that chickens are contaminated with industrial compounds such as PCBs; that additives by the hundreds have never been tested, and that when they *are*, some prove to be hazardous? In the light of the facts, some governmental reassurances are beginning to take on the appearance of the Emperor's new clothes.

To counteract the public reaction, industry has stepped up its public-information programs, such as the one launched by the Institute of Food Technologists to give the press and the communications media the "straight" story on such con troversial food topics as the significance of mercury in food, the use of nitrites and nitrates, and the nutritional changes in processed food. The IFT has set up "expert committees" and fourteen regional representatives whom the press can call on for the industry point of view in a crisis. Included on the panels is Dr. Darby, Dr. Emil Mrak and Arthur T. Schramm –all known for their permissive attitudes toward chemicals in foods. Dr. Darby and Schramm have long griped about the lack of a "balanced" point of view in the lay press when it comes to food additives.

Right after the cyclamate ban, Schramm, in his capacity as chairman of the National Academy of Sciences' Industry Liaison Committee, wrote a "Dear Bill" letter to Darby which some scientists considered at the least indelicate considering Darby's position as chairman of the Food Protection Committee. Schramm complained that

> recent developments concerning the safety of certain GRAS chemicals have put the entire food processing industry in an embarrassingly defensive position;—one that is not warranted by the objective scientific facts.
>
> The communications media, particularly TV, have given considerable publicity to experimental data actually having little or no bearing on safety, but presented in such manner as to dispose a large majority of the lay public to draw dire conclusions. I am referring, of course, to TV reports on chick embryo studies involving cyclamates and subcutaneous injection studies involving sodium glutamate [MSG]. Both reports were inconclusive with respect to the safety of their use in food, but this type of psychological pressure has already taken its toll in the defensive action recently employed by the baby food manufacturers.
>
> The entire atmosphere growing out of such TV programming, coupled with politically oriented Congressional hearings and careless statements by apparently qualified publicity seeking individuals, is one of economic terrorism. Many of the members of the Industrial Liaison Panel, recognizing this sinister development, have expressed strong feelings on the subject and have asked me, as chairman of the Industry Committee, to urge you, as chairman of the Food Protection Committee, to take steps necessary to secure equal time for qualified members of the scientific community to put this matter in proper perspective for the public.

In other words, Schramm was pressuring Darby to make w:
for an industry point of view.

The FDA, for its part, continues to minimize the dang
of food additives. In an interview with *FDA Papers* (no
called *FDA Consumer*) of March 1972, Dr. Wodicka w
asked, "Is the American food supply safe at the present time"
Answer: "It is safer than it has ever been before, and as sa
as we know how to make it . . ."

Dr. Wodicka went on to list the priority concerns of tl
FDA about food safety. In order, they were foodborne disea
such as botulism and aflatoxin; malnutrition; environmen'
contaminants such as industrial chemicals; naturally occurri
toxins—that is, natural poisons in vegetables, such as cabbag
lima beans and potatoes, and in certain seafoods; pestici
residues; and, sixth, food additives. In Dr. Wodicka's vie
any hazard from food chemicals is the lowest rung on tl
ladder; preceding it are any number of problems in whi
food processors have the least direct implication—malnut
tion, environmental contaminants, natural toxins. The on
item on the list for which food processors are directly respon
ble, and which is most within their power and the FDA's
correct, is food additives. Depressing it to a low priority, ho
ever, diminishes the need for action.

Even the toxins put into our food by Mother Nature ala>
the FDA more than man-made chemical additives. Of la
the FDA has begun to focus on ferreting out natural hazai
in common foods such as vegetables. It is true that cabba
may contain a goiter-promoting chemical; spinach may cc
tain nitrate (which may explain why kids spit it out—th
may be smarter than adults); potatoes, especially wh
sprouting or blighted, contain solanine, which has long be
known to be toxic. According to studies there are hundre
of other natural toxins in the food we consider the safest. A
Dr. Julius Coon, who has made a study of the subject, cc
tends that our ordinary food is so filled with toxins that
keep them "in balance"—so that they neutralize each otl
or so that we don't consume one toxin in great quantity—
should not eat too much of a single food but get our vitami
and minerals from a wide variety of sources. Strangely, soi
proponents of food additives use the same argument, sayi

that the more additives we use the better, so that we don't consume great quantities of any one which could be toxic.

Scientifically, of course, there is no reason to exonerate age-old foods from a poisoning capacity. And people who hark back to natural food as "pure" and "good for us" in reaction against synthetic foods may be as mistaken as those who say all additives are perfectly safe. After all, birth defects and cancer and disease have plagued man since the Creation, and some of the trouble may literally come from the apple.

Natural foods, like many chemicals, simply have not been tested for long-term toxicity and should be studied, since all chemicals, even natural ones, may have some risk. However, it is folly to concentrate on natural foods as public policy while ignoring potential hazards from synthetic food additives. Moreover, if natural foods are filled with toxins, of which we know little, why compound the danger by adding synthetic chemicals to foods?

Even when legally compelled to take action on food chemicals, the FDA downplays it, minimizing the hazard to such an extent that the public is bound to wonder why it was necessary. In banning cyclamates, Secretary Robert Finch apologized at his press conference, "I have acted under the provisions of the law because . . . I am required to do so." Other HEW officials let it be known they thought enforcing this law was a lot of poppycock.

After mercury, lead and arsenic were found in fish, beef, shellfish and chicken livers, Surgeon General Jesse R. Steinfeld testified that these were extraordinary findings in the country's generally safe food supply and insisted that "the country is not presently faced with widespread, serious human health hazards." Senator Philip Hart of Michigan, on the other hand, said he found the news some of "the most depressing" he had "heard in quite some time."

When brominated vegetable oils were removed from the GRAS list because they caused heart lesions in animals, the FDA announced that the oils were not "considered an imminent hazard to humans."

After FDA officials found PCBs, considered more acutely toxic than DDT, in fish, cheese, milk and eggs on grocery shelves, and estimated that half the human population carries

PCBs in their tissues, Dr. Albert Kolbye, Jr., deputy director of the FDA's Bureau of Foods, said in a speech in New York that despite their widespread presence, the chemicals were not considered "a threat to human health." "I think the problem is under control," he said.

When forced to ban from feed the cattle-fattening hormone DES, Commissioner Charles Edwards insisted: "DES has been used in the feed of cattle and sheep for nearly two decades without a single known instance of human harm." Edwards stressed that DES "is not a public health problem," but that it had to be banned anyway under the Delaney clause. Like Finch's before him, Edwards' voice was full of regret.

The most common rotten chestnut thrown into nearly every FDA press release announcing action on a food additive is that "there is absolutely no evidence that humans have been harmed by this chemical." But it is never followed up by the fact that nobody has looked to see whether humans have been harmed or that there is absolutely no way of determining such harm, since isolating a cause–effect relationship when millions of people are eating thousands of chemicals is a needle-in-the-haystack proposition. Thus this statement, as every scientist knows and every official at FDA should know, is meaningless and serves no purpose except to deceive an unknowing public into believing there is, therefore, no reason to be concerned.

Similarly common in government terminology is "There is no imminent or immediate hazard to health" or "no public-health danger," followed by "therefore, chemical X will remain on the market until January 1" of whatever year. What does "imminent hazard to health" mean, anyway? If it means we did not find chemical X had arsenical properties that will kill three fourths of the population who eat it in two days, then there is no "imminent hazard." But if it means that every day the chemical continues in existence tissue cells may be harmed, then there is "imminent" (or, as the dictionary defines it, "impending") danger. Perhaps not coincidentally, the "imminent hazard" has an incubation period that usually matches the time it takes industry to use up its stocks and get rolling with newly formulated products.

Obviously, a hazard is a hazard—whether you eat it on July 1 or six months later on January 1. Can it matter to a mother who has a child born with no arms whether she ate Red 2, a teratogenic food dye, during the period when it was not governmentally classified as an "imminent hazard"? To her it was an imminent hazard. And to someone whose system is overloaded with chemical carcinogens, that extra amount consumed in the grace period may be "an imminent hazard." The concept is based on that outmoded philosophy of acute toxicity. The consumer cannot be harmed by immediately cutting off the sale of food additives when they are declared unsafe.

If it seems harsh to lower the boom on industry immediately, instead of granting prolonged recuperation periods, remember several things: 1. Industry is perfectly aware of its legal responsibility as the makers and users of food chemicals to prove them safe. 2. Most chemicals are produced by large companies which are knowledgeable and have Washington lobbyists and lawyers galore. 3. There are usually warnings of danger from preceding acute or subchronic or other tests, all of which lead up to the final conclusion. Sometimes studies are not even undertaken unless there is suggestion of hazard in the first place. Thus industry is usually forewarned; if it fails to heed the warnings we should not be forced to subsidize its irresponsibility with our bodies—nor for that matter with our tax money through loss compensation, as some have suggested.

Remember, too, that the industry is permitted to use these additives on the presumption that they are safe. As Representative Delaney commented recently, "The nation's health and that of future generations demand that regulatory agencies immediately stop the practice of vesting chemicals with rights. Chemicals do not have rights: people do. Chemicals generally are highly toxic and should be considered guilty of being hazardous until proved innocuous."

Since industry reaps the profits from food additives, it should take the risks. The situation should not be turned topsy-turvy so that we assume the health risks for the economic gain of the food producers. It is not only their legal but their moral responsibility to conduct rigorous tests for safety and to anticipate dangers that may arise. Since so few additives have been tested, industry should not be surprised when they turn out harmful, and it should be prepared to take the consequences.

The Anti-Delaney Crusade

The Delaney clause states: "No additive shall be deemed to be safe if it is found to induce cancer when ingested by man or animal or if it is found, after tests which are appropriate for the evaluation of the safety of food additives, to induce cancer in man or animal . . ." It is our most ironclad protection. Naturally, because the Delaney clause robs industry of its customary influence on FDA decisions, it is opposed by industry, which has been recently joined by top officials at the FDA, on President Nixon's staff and in the parent department, HEW.

In March 1972, Foster D. Snell, Inc., Biological Science Laboratories, sent to a number of scientists—including myself —a questionnaire to gather opinion on the dangers of saccharin and whether or not the Delaney clause should be repealed. The solicitation did not appear objective and bore the stamp of an anti-Delaney-clause campaign. Enclosed was the statement of an anonymous physician; a few excerpts will illustrate the quality of his bias:

> Last week it was announced that saccharin was to be removed from the GRAS (Generally Recognized As Safe) list. This makes it susceptible to action under the Delaney Amendment, which means that if one laboratory animal under ANY set of circumstances, no matter how bizarre, can be shown to develop a malignant tumor, then that product, by law, must be summarily removed from the marketplace, no matter how useful or critical it is to our society.
>
> At the present time a few rats in the Wisconsin Alumni Research Foundation study show signs of bladder tumors as a result of being given dosages of saccharin far beyond the range of reasonable scientific investigation. The two-year-old rats (equal to a human's 90 years) were fed the equivalent in human consumption of 875 bottles of diet soda per day for a lifetime. If any one of these tumors proves to be malignant, then by law,

the last remaining non-nutritive sugar substitute will be removed from our culture.

Numerous inquiries by Dr. Samuel Epstein to Snell failed to determine who had asked for the survey, who was paying for it and for what purpose it was to be used. Finally, *Food Chemical News* reported that the whole thing was paid for by an "ad hoc committee [comprised of unnamed physicians and firms] for the fair evaluation of saccharin," and that the physician whose views were expressed was Dr. Robert C. Atkins of New York City (author of *Dr. Atkins' Diet Revolution*). Moreover, *Food Chemical News* reported speculation that the FDA was, in fact, behind the poll.[4]

Ever since the banning of cyclamate, government and industry, almost in unison, have been quietly hammering away at the Delaney clause, trying to think of politically acceptable ways to modify it, which *per se* would destroy it. Pressures mounted again against the law in 1972 when the government was compelled to ban the carcinogen DES, which when put into cattle feed left residues in meat. During the controversy, one of President Nixon's advisers declared that the Delaney clause would have to be "reviewed." In September 1972, before Senator Nelson's subcommittee, Commissioner Edwards devoted most of his testimony to presenting arguments on why the Delaney clause was scientifically invalid. In October in a Denver speech to physicians he again outlined the arguments and deplored the action against DES, "whose benefits to society in terms of food production and nutrition have to be weighed against its theoretical risks as a carcinogen." The Delaney clause, he noted, made a weighing of that risk–benefit impossible. He continued:

> The Delaney Amendment represents one extreme in the continuing search for accommodation between benefit and risk. Unfortunately, the Delaney Amendment is at the extreme of *no* accommodation.
>
> It fails to accept that in fourteen years since its passage, technology to measure in parts per billion has at time outstripped our ability to interpret the meaning of such findings.
>
> We are approaching the capability of detecting the presence of single molecules of materials under study. Yet, even if it were possible to show that a single molecule of a carcinogen pro-

duced no effect in animals or man, that material would still have to be removed from the food supply. And if the Delaney Amendment—as some are suggesting—were broadened to cover teragenicity, mutagenicity and chronic biological injury, the dilemma we would face would be immense.

It seems to me that time and technology have brought us to the point where Congress and the scientific world must re-evaluate some of its thinking about regulatory law. It is simply not enough to hand down legal dictates that allow no measure of scientific judgment and reflect no limit on human abilities to achieve absolute safety in anything.[5]

The battle lines on the Delaney clause are now rather firmly drawn. On one side are certain high-ranking government officials—the "policy-makers"—and industry, including a number of scientists working in private laboratories under contract for industry. For example, at the 1970 White House Conference on Food, Nutrition and Health, Dr. Leon Golberg of the Institute of Experimental Pathology and Toxicology, Albany Medical College, gave an impassioned plea for repealing the Delaney clause and received a rousing hand of applause from industry representatives.

On the other side, supporting the Delaney clause, are consumer advocates such as James Turner and Ralph Nader's Health Research Group, a number of pro-consumer members of Congress including Senator Nelson, and—perhaps most importantly—the whole weight of the most respected scientific opinion in the world on chemical carcinogenesis, including the National Cancer Institute. As Dr. Epstein, in testimony before Senator Nelson's committee, has noted, "It is striking that no such criticisms [of the Delaney clause] have emerged from qualified independent experts, from the scientific staff of the National Cancer Institute, from the membership of the International Union Against Cancer, from the American Cancer Society, or from qualified scientific representatives of citizen, consumer and public-interest groups." All of the clamor to change the Delaney clause, he points out, comes from "captive" industry spokesmen, distinguished in their "lack of expertise and lack of national recognition in the field of chemical carcinogenesis."[6]

Dr. Epstein's contention was upheld in January 1973 when

the prestigious New York Academy of Sciences, noting increased activity to destroy the Delaney clause, called together more than a hundred prominent scientists, lawyers and executives from government agencies, industry and academia to discuss the issue. The consensus was that there was *no justification* for adding a carcinogen to food regardless of the so-called benefits. When asked, all twelve participants chosen by the attending press to answer questions agreed they could not think of a single scientist outside industry who openly favored a change in the Delaney clause.

It's important to note that the FDA was conspicuously absent from this conference, despite repeated invitations to attend. The commissioner sent no representative, and reportedly half a dozen staffers who had accepted invitations to participate hastily withdrew. Thus, FDA officials openly displayed their bias against a law they were hired to uphold.

In contrast, the day after the New York meeting, Dr. Wodicka reportedly went to Florida to confer with industry representatives who had at Dr. Edwards' suggestion formed a "Citizens' Commission" to set up an international symposium to study the broad question of "benefit–risk" of food additives. Dr. Edwards, irked by the Delaney clause, had urged such an international symposium after his September testimony to Senator Nelson's committee; and soon afterward the FDA began working with industry groups to set up the ill-named Citizens' Commission. It is composed of and financed by the Rockefeller Foundation, the Macy Foundation, the National Science Board, the Nutrition Foundation and the Food and Drug Law Institute, and its express purpose is to destroy the Delaney clause. Consumers' Union when invited to participate flatly refused.

Both industry and the FDA try to make the swirl of argument around the Delaney clause sound scientific rather than economic, fearing that the latter would be blatantly offensive to the public. Thus opponents, including the FDA, have drawn up a list of "scientific objections" to the clause—objections which in top scientific circles are viewed in the most charitable language as ill-informed.

Here are the arguments you'll hear time and again from

the FDA and industry scientists in attacking the Delaney clause, and the reasons why top cancer experts disagree:

1. *It is unnecessary to ban all cancer-causing chemicals, as the law demands, because we can set safe low-level amounts of carcinogens for humans.*

"Garbage," is the scientific retort. This argument denies everything we know about carcinogenic effects, essentially that they are cumulative. Therefore, who gives a hang even if we could establish a so-called safe dose for a single cancer-causing chemical (which is unlikely), because such a "safe" dose would be meaningless in the context of our lives. We would still be taking in small doses of carcinogens from various sources: a whiff of polluted air, a bit of contaminated food, the inhalation of cigarette smoke, all of which would add up to make a so-called safe dose a lethal dose.

Dr. Umberto Saffiotti, a supporter of the Delaney clause, has repeatedly pointed out that it is impossible to set such a "safe" level because no one knows how little of a carcinogen it takes to trigger cancer at the cellular level, and the effects of small levels of carcinogens we are exposed to are unknown and cumulative, and potentiating to each other. "We have produced no evidence there is a safe dose of a carcinogen," he declared. "Exposure to cancer-producing agents . . . shows that the tissue which is exposed acquires *new* susceptibility. The tissue stores the damage, so to speak, so that even with low levels of a carcinogen if the cells are exposed to additional insults, the damage can add up so you reach a level where a tumor develops. All of this gives us a situation in which we cannot assume that low level exposure to carcinogens may be insignificant."

His viewpoint, which is also that of his colleagues at the National Cancer Institute, is shared by the weight of scientific opinion throughout the world. A report from a committee of international scientists for the World Health Organization is unequivocal in its opposition to setting any safe doses for carcinogens. "The uncertainty of the extrapolation of the safe dose to man, and lack of knowledge of the possible summating or potentiating effects of different carcinogens in the total human environment preclude the establishment of a safe dose at the present time on grounds of prudence."[7] Even certain scientists who hold the opposite view—that eating small amounts of carcinogens is not dangerous—such as Dr. Leon Golberg, admit that they are "in the minority." Yet it is this minority, instead of the National Cancer Institute, that the FDA listens to.

2. *The Delaney clause is invalid because any chemical, administered in high enough doses, will produce cancer in animals.*

Dr. Golberg has written: "It is safe to predict that, by appropriate choice of dose, concentration of solution and frequency of administration by the subcutaneous route, any chemical agent can be shown to be a carcinogen in the rat and probably also in other species of laboratory rodent." Although it is true that subcutaneous injections of chemicals do frequently cause cancer, especially local lesions, it is unusual for chemicals *ingested* by animals to cause cancer, and the Delaney clause specifies ingestion as the route of administration. In a study by Bionetics Research Laboratories done in 1970, mice were fed 120 compounds at the *highest* doses they could tolerate without dying. Less than 10 percent of the chemicals produced cancer.[8]

3. *Our scientific methods of detecting scant amounts of carcinogens are now becoming so sophisticated that if we ban everything that contains even a molecule of a carcinogen, we'll starve.*

As other arguments are discredited, this one is being increasingly put forth to obfuscate the issue, and at first blush it seems to have merit. Dr. Edwards used it extensively in his September 1972 Senate testimony. He said in effect that the concept of zero tolerance and zero residues of cancer-causing substances would make mandatory the banning of "all animal and plant products containing not only detectable levels of naturally occurring carcinogens, as well as those having the potential to contain a few molecules of a naturally occurring carcinogen, but also all food containing such carcinogenic environmental contaminants as traces of radioactive materials."

This line of argument, which implies that we would have to ban virtually all foods, is an enormous red herring. It is true that advancing methodology makes it possible to detect traces of carcinogens in foods at only one part per billion. However, Dr. Edwards' argument applies mainly to *accidental* residues, such as pesticides, that may *inadvertently* and without our knowledge get into food and that as such do not come under the Delaney clause anyway. Legal interpretations of the clause made by the FDA itself make it applicable only to direct or indirect additives in food or in animal feed which may leave residues in food.

One might think from the FDA's protestations that it had had to invoke the Delaney clause at every turn as science turned up new methods of detecting low levels of carcinogens. Such is hardly the case. According to FDA general counsel, Peter B. Hutt, in the fifteen years since the amendment was passed it has been invoked only twice![9]

4. The Delaney clause is unscientifically restrictive because it allows no room for scientific judgment.

Industry spokesmen insist that if cancer shows up in only one rat fed a certain chemical, no matter how sleazy or invalid the experiment, the chemical would have to be banned. This is untrue. As Dr. Saffiotti, among other scientists, points out, there is plenty of room for scientific judgment, because the clause says that the chemical must have *caused* cancer. To determine if it did cause cancer, expert scientists like Dr. Saffiotti must be called upon to make the judgment, and to do so they must determine if the test methods were sound.

The Pine Bluff Experiment

In January 1971, Dale R. Lindsay, the FDA's Associate Commissioner for Science, told the Nutrition Foundation:

> We have indulged ourselves in a luxury which we cannot much longer afford, that is, to arbitrarily ban consumer products because they contain inadvertent traces of chemicals which have been shown to produce one or another of the adverse effects just named [cancer, birth defects, mutations] irrespective of the size or the method of administration of the dosage given. Some of these bans are warranted beyond any doubt; others are doubtful. We must find out how to calibrate our hazards with dosage in experimental animals and then how to extrapolate our findings to man.

In order to do this, Lindsay said, the FDA intends "to seek a better working relationship with the scientists in industry." Although the FDA is "painfully aware of the dangers we face from today's rampant consumerism in attempting to develop closer relationships with industry," he declared, "we are convinced that it must be done."[10]

Dale Lindsay is now chairman of the scientific advisory board for a large new testing laboratory operated jointly by the FDA and the Environmental Protection Agency (EPA) at what used to be an army arsenal in Pine Bluff, Arkansas. Called the National Center for Toxicological Research, its purpose, according to FDA, is not to test out the toxicity of certain substances, but to conduct theoretical investigations into the nature of toxicology itself, such as exploring dose–response relationships and comparative toxicology among species. True to his word, Lindsay has drawn advice from the Society of Toxicology, which is dominated by industry scientists, and his official board of advisers is made up of numerous industry personnel and unofficial industry spokesmen such as Dr. Golberg.

Perhaps not surprisingly, then, the plans for Pine Bluff reflect industry prejudices. The goal of the Center's very first project, which will take years, is to prove that carcinogens, if tested in enough animals, do have "no effect" levels and that thus there is a base below which carcinogens can be used in foods without causing cancer. The FDA hopes to show this by running "megamouse" experiments in which thousands of animals are tested at low doses with known carcinogens. As Dr. Albert Kolbye, deputy director of the Bureau of Foods, has said, "One of the major thrusts of the National Center will be to test large populations of animals with relatively low-level, long-term exposures to chemicals that have been currently identified as being potentially carcinogenic, such as DDT." In explaining why the FDA has given this project priority, he said, "We have good reasons to suspect that man can tolerate exposure to certain substances without adverse effect ensuing. By this I do not mean to imply that we are going to abrogate our responsibility to protect public health, nor do I mean to challenge some of the good intentions behind the Delaney amendment. However, we must face the reality in that there are many potential carcinogenic substances naturally found in our food supply for which we must develop a meaningful regulatory posture."

Subsequently, for Phase I of the Center's activities, a scientific task force headed by Dr. Golberg in May 1973 began testing three known bladder carcinogens (2-acetylaminofluorine, N-hydroxy-2-acetylaminofluorine and 4-ethylsulfonylnaphthalene-1-sulfonamide) on large numbers of two strains of mice.

Clearly, this indicates that under the guise of basic research the FDA is using your tax money—quite a bit of it—to try to prove a pet theory that carcinogens can be used safely in food, and to subvert the Delaney clause. The experiments will be used, then, to decide not which chemicals are carcinogens and unsafe for you to eat, but *how much* of a carcinogen you should be allowed to eat. The FDA is saying, Let's use 50,000 or 100,000 animals—instead of the usual fifty—and see if we can pick up the risk at a small percent, perhaps even a tenth of a percent. (Of course, a tenth of a percent, one rat in 10,000, is 20,000 Americans who may still think the risk of cancer intolerable, even at low levels.)

In one sense, using large numbers of animals could be beneficial because we could detect weak carcinogens that go unnoticed in small-number experiments. However, the whole approach is senseless and wasteful when there are so many chemicals that may be *strong* carcinogens that have never been tested. The chemicals to be studied at Pine Bluff, remember, are *known carcinogens* for which they are seeking a "safe" dose level. But even if they could establish a "safe" dose for a single chemical, it would be meaningless, since there are so many carcinogens loose in the environment and the effects are cumulative.

Dr. Saffiotti privately and publicly has advised the FDA of the fallacy of the experiments, and stated his view before the Subcommittee on Executive Reorganization and Government Research of the Senate Committee on Government Operations:

> I personally believe that certain approaches to the problem of identifying a "safe threshold" for carcinogens are scientifically

and economically unsound. I have in mind some proposals to test graded doses of one carcinogen, down to extremely low levels, such as those to which a human population may be exposed through, say, residues in food. In order to detect possible low incidences of tumors, such a study would use large numbers of mice, of the order of magnitude of 100,000 mice per experiment ["megamouse" experiment].

This approach seems to assume that such a study would reveal that there is a threshold dose below which the carcinogen is no longer effective, and therefore that a "safe dose" can be identified in this manner. Now, there is presently no scientific basis for assuming that such a threshold would appear. Chances are that such a "megamouse experiment" would actually confirm that no threshold can be determined. But let us assume that the results showed a lack of measurable tumor response below a certain dose level in the selected set of experimental conditions and for the single carcinogen under test. In order to base any generalization for safety extrapolations on such a hypothetical finding, one would have to confirm it and extend it to include other carcinogens and other experimental conditions such as variations in diet, in vehicles used, in the age of the animals, their sex, etc. Each of these tests would then imply other megamouse experiments.

Dr. Saffiotti went on to point out that "megamouse" experiments are a bad bargain. We can spend, he said, $300 million to test twenty chemicals each on 100,000 mice; whereas for the same amount we can test four thousand chemicals each on five hundred mice, thereby filtering out our strongest environmental carcinogens, from which we can then protect ourselves.[11]

It is somewhat astounding that with the best scientific minds available under the HEW umbrella, of which the National Cancer Institute is part, the FDA shuns such advice and relies instead on the self-serving scientific expertise of industry.

As long as the FDA and industry are of such like mind, we don't need a Ouija board to predict that in crises over food additives we will be the losers.

II

Five Case Histories of Nonprotection: Proof Positive

Many highly sophisticated and intelligent people truly believe that the government's decisions about food additives
are made rationally on a pure scientific and legal basis in the
best interests of the public health. In fact such decisions are
largely political and economic, and when science and the
public interest win out it is invariably only after the government has been pushed to the wall by consumer advocates
and other public pressure. It is hard to imagine the day-to-day
battles over irrelevant points, the evasions, half-truths, deliberately misleading statements and corruption of scientific data
which official statements feed to the public as reasons for
action or nonaction. Part of the problem obviously is immov-
able bureaucracy, inertia, a resistance to change in the status
quo, the defensive posture that power breeds, and a mistaken
view of one's "public service."

In Part I of this book we've tried to give a background for
understanding how your protection from food hazards is un-
dermined. In Part II we offer "proof positive" of protection
you aren't getting: five detailed case histories of how the government has reacted when faced with major decisions about
food-additive hazards.

You may wonder, as you read the cases, how could it happen? How could government officials refuse to act on the
basis of such compelling evidence? But, if the facts seem

shocking, it is because they *are* shocking. There is no other explanation; it is futile to try to pin governmental food decisions on a base of pure science.

For one thing, the working objective scientists at the FDA comprise a relatively small group at a low echelon of influence; they generate data on which decisions are to be made, but only rarely are they given an opportunity to recommend specific policy (though scientific expertise was more welcome in former years than today). Their data are passed through layers of bureaucratic middlemen, who interpret them, sometimes launder or distort them, and, though some have some scientific background, often become adjunct commissioners more concerned with pleasing superiors and advancing their careers than sticking to their scientific guns.

The decision-makers are rarely working scientists or specialists in areas in which they make decisions; they are management experts (Dr. Edwards, who is a surgeon, made his reputation as a management whiz before coming to the FDA) at the top of the bureaucratic ladder, where they are subject to political and economic pressures. The top positions in the FDA and HEW are political appointments of the President. In some instances, it seems clear that decisions are not based on the agency's scientific expertise, but are imposed from *above* by the HEW hierarchy or the White House, sometimes over objections of some at the FDA. The thwarted effort to keep cyclamated foods on the market, for example, by having them declared drugs was a policy that Commissioner Ley said "came downward from the Secretary level instead of a recommendation going upward from FDA." There have been numerous instances of topside administration interference with FDA decisions, especially in the drug field—so much so that Senator Warren G. Magnuson termed the FDA's science "politico-science, where expediency, the status quo and private interests receive more consideration than scientific truth and public good." In some instances, as you will see, the scientific justification seems so screwy that it appears certain the decision was made for economic-political reasons and then science twisted to fit around it.

Under such assaults, science and consumer protection at the FDA are sliding downhill rapidly, despite official protestations to the contrary. Dr. Edwards has said he believes that the scientific strength of the agency is the highest in years. He said in 1973, "A regulatory agency like FDA must have scientific credibility; I believe that FDA today has that credibility."[1]

To be sure, scientific decisions are rarely cut and dried. Judgments enter in, points of view, backgrounds, ambitions. Scientists have been asked to conceal or alter information hostile to the use of additives; memos have been changed without scientists' knowledge to prevent unwanted opinion from reaching supervisors. Scientists have made mistakes in interpreting data and failed to look for hazards in areas we now know are important. But in retrospect these seem minor transgressions.

Today there is a sensing that within the agency there is a much more invidious force at work: a move to degrade science as a matter of course. Much of the agency's dissenting vitality has left—scientists like Dr. Marvin Legator, Dr. Herman Kraybill, Dr. Kent Davis, who could be counted on to strongly disagree with "establishment" policy when it was scientifically untenable. The FDA does not, in the opinion of some authorities since Dr. Kraybill's departure to the National Cancer Institute, have a single topflight authority on chemical carcinogenesis, the area in which the FDA makes most of its decisions. This lack of expertise is compounded by the FDA's increasingly turning toward a small clique of pro-industry scientists and away from renowned scientific authorities. Also, often there is no genuine disagreement about the scientific facts, only about how to use them. For example, every one, including the FDA hierarchy, may agree that a certain food chemical is a teratogen in rats or in chicks; where the disagreement comes in is on how stringently that knowledge should be applied to humans.

Many FDA officials undoubtedly believe they are doing a good job of protecting consumers. Some FDA'ers feel truly buffeted by pressures from industry, rationalizing that they have taken a "middle posture," disliked by industry and "radical consumer advocates" alike. As one official says, "We're not really out to do in the consumer, you know." But this so-called "middle road" is deceptive, for, while there is what Harrison

Wellford calls a "consumer advocacy gap," industry, with its battalion of lobbyists, is adept at making it seem that the facts are compelling on their side. The officials come to believe that the interests of the agency, the consumer and the food industry are one and indistinguishable.

This kind of thinking is bound to create an atmosphere in which the agency becomes the defender of the status quo. Instead of admitting the existence of real problems which need correction, officials hostile to change and afraid of criticism deny that anything is amiss. This leads to a defensive stick-togetherness. A word against an agency decision is sometimes interpreted as criticism of everyone in it. Some scientists when they get into the inner circles begin to "see the other side." Commissioner Edwards, according to an analysis by Jim Turner, came to the FDA with promise and gusto, but very quickly became enmeshed in routine inbred bureaucratic thinking, and his resolve eroded.[2]

This background is given not to excuse the FDA's conduct, but to help make it more comprehensible. Also, it must be said that there are many dedicated competent persons in the FDA who are unhappy about many of the FDA's official decisions.

The important point to remember as you read the following cases is that in decision-making at the FDA, faithfulness to science is tempered by politics and the explosive economic impact on the richest industry in America: the food industry. Though such decisions may infuriate you, they are not the result of inexplicable villainy, but rather of human shortsightedness, external pressures, personal ambition, lack of courage and lack of foresight in anticipating the human consequences.

Red 2: The Abortion Pill You May Not Want

You might have thought that the FDA had no forewarning that FD&C Red No. 2, the official name for our most widely used food dye, which gets into practically everything, could possibly be dangerous. That was the official "reaction" when it was learned in 1970 that Russian scientists had published two new studies incriminating amaranth, another name for the dye. One Russian study showed that the red dye caused cancer in rats; the other was the first to show that the dye caused birth defects, stillbirths, sterility and early fetal deaths in rats given the dye in exceedingly small amounts.[1]

In fact, the Russians found a danger to rat fetuses that were fed only 1.5 milligrams per kilogram of body weight, which is the *exact* dose established by the World Health Organization as the "safe dose" for humans—the acceptable daily intake, or ADI, as it is called. In other words, there was no margin of safety at all. Rats eating precisely the same amount of Red 2 as we do per kilogram of body weight were "resorbing" the fetuses early in pregnancy, comparable in humans to a miscarriage, or spontaneous abortion.

The Russians did not stumble onto the Red-2 hazard blindly. They were merely following the advice of worldwide scientific opinion to further test out Red 2—advice which we should also have taken, but hadn't. As long ago as 1956 at a Rome symposium of scientists, amaranth was classified as a compound suspected of being carcinogenic. And in 1964 a joint FAO/WHO Expert Committee on Food Additives cleared Red 2 for food use only with the caution that it undergo immediate further study. Soon afterward the Russians banned amaranth paste in food after finding it carcinogenic in rats. Searching for a substitute, they tested pure amaranth and came up with their new cancer data, which reached the FDA's ears in late 1970 when translations became available.

But the Russian cancer study didn't cause much of a stir at the FDA, because, in one official's words, it was "old hat." It was the second Russian study on reproductive danger from the red dye that caught the FDA's official attention. This was indeed new stuff, for though studies on reproductive danger had been urged by the 1966 Expert Committee on Food Additives, no one had made them. Specifically, the Russians found that amaranth given to female albino rats for twelve to fourteen months caused a drop in fertility and an increase in stillbirths and deformed fetuses. It was an alarming finding to FDA scientists, for we knew that the whole population was getting this chemical on a daily basis—and of course that included millions of pregnant women.

It is important to point out that Red 2 is almost impossible to avoid, because it is in virtually all foods, not only in those obviously colored red, and it is cheap and easy to use. In an average year, according to James E. Noonan of the Warner-Jenkinson Company, one of the producers of Red 2, about $19 million worth of the chemical is produced and put into an estimated $15 billion to $25 billion worth of food. One FDA official has said that if everything containing Red 2 self-destructed, a lot of people would starve to death.[2]

In its pure form, Red 2 is a reddish-brown powder which in solutions turns a deep red that is perfect, say food technologists, for producing attractive brilliant hues—for example, a grape shade when mixed with blue. Alone and in combination with other dyes, it goes to simulate color associated with the flavors of caramel, apricot, strawberry, raspberry, cherry, currant, blackberry, grape, chocolate. Red 2, not cocoa, is often responsible for that dark, rich look of chocolate cakes and brownies. On the other hand, it is sometimes used to lend "that precise shade" to white icing, as well as the bright red in hot-dog casings.

Here's just a partial list of foods containing Red 2, compiled by the FDA in 1971: ice cream, processed cheese, ice milk, luncheon meat, frankfurters, fish fillet (fresh or frozen), shellfish, cornflakes, shredded wheat or wheat cereal, rice flakes or puffed rice, rolls (sweet, cinnamon, Bismarcks, etc.), snack items (pretzels, corn chips, crackers, etc.), cookies, pie crust, cake mix, pickles, peaches, citrus juice and fruit juice (canned), canned apricots, cherries, pears, pineapple, fruit cocktail, salad dressings (French, mayonnaise), jelly, pudding mix, syrup, jam, candy bars, vinegar, noncola soft drinks.

To compound the danger, there's little way to detect Red 2 through labeling; it is noted simply "US color certified added." You can't tell whether the color is Red 2 or another. Red 2 is also used in cosmetics, mainly lipstick, and drugs—including vitamins given to pregnant women.

But when confronted with the new Soviet evidence on Red 2, the FDA promptly dismissed the Russian paper showing carcinogenesis as being worthless because the Russians stated they found no tumors in the control animals, which, the FDA said, was not likely to happen in valid experiments (ordinarily you expect cancer in control animals as they age). Yet there are a lot of data in FDA files in which other experimenters have found no tumors in control animals, and the Russians may have developed a strain of animals highly resistant to cancer. But even if the Russian experiment was not perfect, it did bring to light a hazard that should be taken seriously, and the FDA should have immediately undertaken a cancer experiment to determine whether we would obtain the same results.

Instead, FDA officials insisted to the press that there was no valid suspicion of cancer from Red 2. According to Dr. Wodicka, the dye had been tested thoroughly many times and found "clean." One of these important tests, in fact, was done in the FDA's own laboratories in the early 1950s and has been widely quoted in the literature as supporting the no-cancer danger. But a careful reading of the results in a memo from Dr. A. A. Nelson, then the pathologist who studied the rats, reveals that Nelson did *not* find "no effects." In fact, he found a number of tumors in rats fed Red 2, as well as in rats fed other dyes, Red 4, Yellow 5 and Yellow 6.

But, even more strange, in the process of his analysis he found that seven rats had somehow never been submitted to the pathology lab for detailed dissection and observation. Interestingly, all of these rats had tumors, and six of them had been fed a dye; only one was a control animal. But by that time, conveniently perhaps, there was no way of knowing which of the cancerous animals had eaten which dye, so the poor pathologist could not include these in his evaluation. And this is the kind of study that was disseminated worldwide to support the conclusion that no one had to worry about getting cancer from eating Red 2!

Dr. Nelson's conclusions, according to a memo dated July 2, 1954, were:

> Tumors: The commonest of these were pulmonary lymphosarcoma and breast tumors, the latter occurring only in females and being about equally divided between fibroadenoma and adenocarcinoma. . . . Dr. H. L. Stewart of the National Cancer Institute felt that there was a suspicion that the colors fed were increasing the incidence of the breast tumors. Later on checking Dr. Fitzhugh's [the author of the study] original data, a total of 7 additional subcutaneous (in all probability breast) tumors which for one reason or another had not been sent to the pathology laboratory were found; since 6 of these were in treated rats and 1 in a control the significance of the original suspicion was not lessened.

If the FDA had zeroed in on a possible cancer danger and confirmed it with new studies, Red 2 would have been wiped out overnight under the Delaney clause. Perhaps this explains why the FDA did nothing of the sort. However, finally, in March 1972, it did institute a two-year feeding cancer study on Red 2 at the FDA—with results not due until sometime in 1974. It was an unnecessary delay.

It was much safer for the FDA to pursue the question of birth damage from Red 2, for any detrimental results could be waived at the FDA's discretion—since there are no stringent legal demands for the ban of an agent causing reproductive harm, unlike cancer. Subsequently, in early 1971, the FDA began a program of crash studies which were supposed to confirm or refute the Russian findings on embryo damage from Red 2. In March of that year I began studies on chick embryos, and since the technique gives results within twenty-one days—the time it takes the chicks to hatch—by April I had considerable evidence that the dye was plenty toxic. It caused some deformities, such as skeletal defects, stunted growth, some malformed eyes, but the striking effect was that under certain test conditions so few chicks survived long enough to hatch. They died very early—killed by the dye.

Even more interesting, it took very little of the dye to kill them. At the lowest doses of twenty-five parts per billion, which is an infinitesimal amount, 100 percent of the chicks were killed. Also, at the highest doses 100 percent of the chicks were killed before hatching. There was a seesaw effect: from a certain midpoint, both the more you gave of the dye and the less, the more chicks died. This was not, in toxicological jargon, what they call a "true dose–response relationship." And so, because it did not fit into a typical pattern, the low doses could officially be discounted as not valid, even though this was essentially what the Russians had also found in rats. (In their experiments the effects were greater at 1.5 milligrams, for example, than at fifteen milligrams.)

There may be some perfectly good explanation of why the death rate was higher at the lowest doses than at the medium doses, which killed only 40 percent of the chicks (nothing to take lightly in itself), and that same reaction may occur in people. For example, the body can sometimes more efficiently dispose of a big slug of a chemical without harm because the dose overloads the system and is not completely absorbed and metabolized but is rapidly excreted, whereas smaller doses of the same chemical may stay in the stomach and be completely digested.

In any event, it was not long before my colleague Dr. Thomas Collins in another FDA laboratory came up with similar results in rats. He too found some abnormalities (also not entirely dose-related), and he did find that the fetuses of the rats were dying before birth in significant numbers. Twenty-nine percent of the rat fetuses died at high doses of Red 2, compared with 2.5 percent for plain distilled water. Lower doses (fifteen and 7.5 milligrams per kilogram of body weight daily) also killed the unborn. There seemed no longer any doubt that, as the Russians had found, Red 2 was a poison to developing embryos in rats. And if the FDA believed at all in its tests it would have to conclude that the same thing probably was happening throughout the human population to unsuspecting women.

And by some miracle or miracles the FDA did! In September 1971 it put a notice in the *Federal Register* that it intended as of January 1972—when the provisional listing came up for renewal—to severely cut back on the allowed uses of Red 2. For all practical purposes this edict meant the end of Red 2, for the restrictions were to be so severe as to render it ineffective as a coloring agent and thus useless to the industry. Pink instead of bright-red cherry pop, for example, was not the industry's idea of a viable product. Use of the color would be limited to packaging and other inconsequential or external uses.

Responsibility for this decision lies with the FDA's scientists. In a memo dated November 18, 1971, of a meeting chaired by Dr. Leo Friedman, head of the agency's division of toxicology, the FDA's scientific circle of experts clearly expressed their intent that Red 2 should be banned. Their conclusion was that on the basis of the new FDA studies, the Russian study showing reproductive interference, and the questions raised by the Russians concerning carcinogenesis, "it was agreed that it would be prudent to limit drastically the uses of Red No. 2 only to indirect or incidental applications involving food: that is, limit use of the color to such applications as food packaging where migration to food is nil, color marking on animal feed additives and to external uses only in drugs and cosmetics."

The recommendation was based on simple scientific logic: the hundredfold safety margin in extrapolating animal-test results to humans. When you took the lowest dose at which Red 2 caused birth damage in FDA's rats, 7.5 milligrams, and divided that by one hundred to establish the "safe dose" for humans, it was so small (.075 milligrams per day) that it was ludicrously useless. For example, the daily safe dose for an average person would be far less than the red dye found in one bottle of cherry soda!

With the agency's September ultimatum, FDA scientists felt they had won a major victory. It was indeed a strong regulatory decision, but it didn't last long. Two weeks after the *Federal Register* notice appeared, the industry's Institute of Food Technologists Committee on Nutrition and Food Safety suggested that this was too important an issue to be left to the FDA alone. It called on the FDA to seek an "objective" opinion from an "independent" group such as the National Academy of Sciences.

FDA officials have conceded that there was considerable "industry flak" over the decision to eliminate Red 2. Dr. Wodicka told a reporter, "There were a lot of people with reasonable toxicological credentials who disagreed with the interpretations we at FDA were putting on our own work." He identified the people as industry representatives and outside scientific opinion brought in by industry. As Dr. Wodicka summed it up, there was a "sensing of the climate of opinion that you guys [at the FDA] are going off half-cocked with evidence that does not justify the conclusions."

An attorney for the color industry said there wasn't a day he wasn't on the line with someone at the FDA, pleading his case. Besides an immediate economic impact the industry feared that once the FDA got started curbing a food chemical just because it caused birth defects or abortions—which was unprecedented—Pandora's box would be wide open, exposing all other food chemicals, and particularly colors closely related to Red 2 chemically, to similar fates. In fact, in its September notice the FDA had also ordered industry to get busy on reproductive tests on other food dyes. The industry also undertook its own studies on Red 2; and the FDA, for additional data, contracted with the Food and Drug Research Labs, operated by Dr. Bernard Oser, to test out the reproductive effects of Red 2 on rats, hamsters, rabbits and mice.

Not surprisingly, the industry hammered away at Dr. Collins' rat study in an attempt to discredit it. The study had been done by a method known as "gavage." That is, when the female rat was mated she was also started on doses of Red 2 that was put down into her stomach by tube instead of being fed to her in her diet. This method is exceedingly accurate because you know precisely how big a dose gets into the stomach, and also it mimics the use of red dye in the population, since about 50 percent of the dye is consumed in beverages. To the amusement of some FDA scientists, the industry, in trying to refute Dr. Collins' study, did an experiment by gavage with Red 2 and got even worse results: at a relatively low dose of fifteen milligrams per kilogram a day, 55 percent of the fetuses were resorbed, and at higher doses of 450 and 1,500 milligrams 100 percent of the fetuses did not survive to be born. The industry then came running back to the FDA complaining that the method was unsound. They later came up with other studies, mostly from Red 2 fed in the diet, showing the dye to be harmless or at least less harmful—which became the showcase in the "save Red 2" drama.

By December the FDA had completely given in. The agency's hierarchy overrode its scientists' advice and turned the matter of Red 2 over to the National Academy of Sciences/National Research Council. From then on, the fate of Red 2 was predictable. The dye was bound to come out tasting as sweet as cyclamate had after having been referred to a similar committee chaired by the same man, back in 1969. Red 2 was not likely to find enemies among the same men who also found that such hazards as saccharin and monosodium glutamate were little to worry about. FDA scientists felt betrayed and thoroughly disgusted, for there hardly has been a case to come up before the FDA that has been more clear-cut. Dr. Marvin Legator, then chief of the Genetics Branch at the FDA and now a professor of genetics at Brown University, aptly characterized the feeling: "If the FDA can't make a decision on this one, they can't make a decision on anything."

At first even Dr. Philip Handler, president of the National Academy, saw the folly of the FDA's seeking a way out of an unpopular, economically explosive decision by putting the burden on the Academy. To the FDA's surprise, Handler initially refused to evaluate Red 2, saying he considered the matter "too routine" and well within the FDA's competence. But Handler later relented, for unknown reasons, and authorized the establishment of an ad-hoc NAS/NRC committee to decide Red 2's fate. It was a move, said Ralph Nader's

Health Research Group, that "diluted" the Academy's prestige and "insulted" the competence of the FDA.

On February 10 and 11, 1972, the ad-hoc committee met in Washington, D.C., to hear testimony from researchers both from FDA and from industry-supported labs. After I presented my material showing the serious defects caused in chicks, Dr. Julius Coon, the chairman, thanked me for "entertaining" them. After studying Dr. Collins' data, one member joked about the apparent abortive effect of Red 2 and suggested it might be turned into a birth-control pill. To some of us it seemed that the committee had its mind made up in advance, that it uncritically accepted data which claimed Red 2 presented no hazard and spent most of its time trying to discredit the findings of the FDA which showed Red 2 to be dangerous.

Following the hearings (they are closed to the public, and no record of the proceedings is made) the committee went into executive session to consider the various studies: the FDA data; an industry-sponsored feeding study done by Industrial Biotest Laboratories in Illinois which purported to find no danger; another study done for industry by Woodard Labs in Virginia which did show danger by gavage but not by diet; studies by the Food and Drug Research Laboratories in New York which also purported to show no danger; and a mutagenesis study done under contract for the FDA which did not show danger from mutations. Reportedly, the committee's report and recommendations were written the following day—and when news of the recommendations reached the FDA laboratories soon afterward, even those who had expected the worst were surprised. It was a whitewash (or, as I have often put it, a "redwash"). One toxicologist remarked how appropriate it was that a snowstorm blanketed the FDA building white on the day we received the news.

In clearing Red 2 the committee relied on such old, deceptive chestnuts as that it had been "in widespread use since the early years of this century without suggestion of harmful effect on human health"—as if that made an iota of difference —and that there were "inconsistencies" between some reports showing hazard and others not showing hazard. Furthermore, the committee's report stated, none of the evidence was "so conclusive or convincing that it [could] be extrapolated to health hazard in adults, pregnant women, or children." In other words, the committee even threw out the window the whole basis for the tests—that they had relevance to humans. The committee said that it appreciated "the concern that led the Food and Drug Administration to its proposed 'allocation of permitted usage patterns' for FD&C Red No. 2" and that it was "sympathetic with the policy of that agency to err on the side of safety." However, the committee concluded that it was "a premature and unnecessary measure at this time" to restrict the use of Red 2 in any way whatever.

When the committee report was submitted to the Academy's review board, however, the panel of prestigious reviewers did not think it worthy of the imprimatur of the Academy. For three months the report was batted back and forth between the reviewers and the authoring committee; but the committee refused to alter its conclusions, and the review panel is advisory only.

In June, somewhat reluctantly—or so it appeared from his covering letter to the FDA—Handler released the report. It was attacked by critics, both for its incompetence—the committee did not include either a teratologist or a geneticist, the very areas on which it was asked to pass judgment—and for its bias. The Health Research Group did an analysis detailing how the raw data presented to the committee did not match the conclusions. *Consumer Reports* pointed out that the committee "curiously," in noting the Woodard Lab study, mentioned only the part that showed no resorptions when Red 2 was put into the diet, and that it pointedly ignored the very impressive detrimental results when the dye was given by gavage.[3]

Dr. Matthew Meselson, professor of biology at Harvard and a member of the Academy itself, confirmed that the FDA reproductive tests performed by Dr. Collins showing evidence of fetal toxicity were "well presented, well designed and appear to have been carefully conducted." He added that, contrary to the Academy committee, he saw no contradiction between FDA studies showing danger and the studies purporting to show no danger. He explains that Dr. Collins started administration of Red 2 at the time of conception—

as did the Russians—whereas other labs did not start feeding it to the animals until they were in their sixth day of pregnancy. This, he believes, may account for the fact that the latter researchers did not see such striking damage.

Unquestionably, in human terms Dr. Collins' approach is more realistic, for a woman consumes Red 2 from the moment of conception; she doesn't just start somewhere in the second trimester of pregnancy. As *Consumer Reports* noted:

> In human terms, the latter method [after six days] is comparable to beginning the dosage near the end of the third month of pregnancy, when major organ systems and biochemical pathways have, in large part, already been formed. If the other experimenters had also started Red 2 at the onset of gestation, their results might well have shown a clearer pattern of fetal toxicity. In any case, the possibility cannot be ignored—particularly in view of the human fetus' well-studied vulnerability in the early months of pregnancy.[4]

Even officials at the FDA were embarrassed by the NAS committee's recommendations and conclusions. Dr. Wodicka had stated previously that if the committee completely cleared Red 2, the FDA could not accept it. The FDA then was in a ticklish situation: it was committed to taking some action, but it could not ignore the Academy's report.

Stuck with the Academy's unacceptable advice, the FDA did an astonishing thing. On the basis of Dr. Collins' rat studies, the FDA set a "safe" level—that is, the level beneath which Red 2 did not cause damage to rat embryos—at fifteen milligrams per kilogram of body weight per day. In short, the FDA accepted its own studies as valid. The proper procedure then would have been to apply the usual hundredfold toxicological-safety margin, putting the safe human intake daily at .15 milligram per kilogram of body weight daily. But as any FDA'er could see, this would still be less than a bottle of cherry soda per day for most people, and the FDA would have been right back with a virtual ban of Red 2, just as FDA scientists had recommended in the fall of 1971. The FDA, however, to quote attorney Anita Johnson, "invented a disarming concept of finding a safety factor which allows the industry to continue to use Red 2—the public be damned."[5]

Simply, the FDA decided to apply a safety margin of *ten instead of one hundred*. Using this tenfold safety margin, by great coincidence, put the "safe" human dose at 1.5 milligrams—precisely what the FDA had been allowing *before* the Red-2 incident blew up. Then the FDA announced it was cutting back on the use of Red 2 by imposing a ceiling of thirty parts per million on the amount allowed in foods and of a thousand parts per million on the amount allowed in drugs and cosmetics. This cutback, it said, was necessary to make it impossible for anyone to consume more than the "safe" level of ninety milligrams per day. Commissioner Edwards billed it as "an additional margin of safety for consumers."

In truth this FDA sleight of hand drastically lessened our protection instead of increasing it, as the ninety-milligram level was based on the slashed safety margin of ten instead of the previously used one hundred. And the upshot? We would be allowed to eat the *same maximum* amounts of Red 2 as we were before the new findings on the hazards of Red 2 ever came up! (The fact that the FDA says it had to cut allowances to get down to that ninety-milligrams-per-day safe usage is tacit admission that the agency had allowed the usage to slip above that proclaimed safe dosage.) Ironically, we would be much worse off than before.

When asked by a reporter for *Medical World News* why the FDA cut the safety margin for Red 2 from one hundred to ten, Dr. Wodicka gave a reply remarkable in its candor: "In the first place we're stuck with Red 2; if we went to a .15 milligram limit we'd wipe out its use."[6] Some might consider this a curious way of approaching a scientific question—to first set up the premise that we must keep Red 2 and then manipulate the scientific data to fit the conclusion. But it is clear that that is what the FDA did. The hundredfold safety margin has rarely been violated by the FDA and is legally required "except where evidence is submitted which justifies the use of a different safety factor," according to federal regulations. Several scientists at the FDA could not recall a single other instance in which official policy had allowed so little as a tenfold safety margin for direct food additives—though a tenfold safety margin has been used for environmental contaminants, such as mercury, which cannot be totally avoided in food.

Officially, the FDA cited two pieces of nonevidence as justification for reducing the safety margin: (1) the National Academy committee's recommendations and (2) safe human experience with the use of the color. Neither of these has any merit. The committee's report is an opinion, not fact, and the "human experience" is a distracting irrelevance, since the gathering of any data on the harm to humans from Red 2 is impossible. Because Americans consume thousands of chemicals, isolating the effects of any particular one through observation is hopeless. You may recall that though thalidomide produced unique deformities and was given to a select population—pregnant women—it still took five years to track down the cause of the deformities. To argue that our protection should be lowered because harmful effects from Red 2 haven't been observed in humans is the grossest kind of dishonesty, meant to deceive a public generally not knowledgeable enough to dispute it.

But what is more remarkable is that, after all the hoopla over its intent to cut back on Red 2, the FDA has never put the proposal into effect. As of April 1973, two years after the FDA became cognizant of the danger, officials confirmed that no action at all had been taken to implement the thirty-parts-per-million proposal in food and beverages. For one thing, industry objected to the ceiling as far too low. Thus the FDA's whirlwind of concern had been all shadow and no substance. And once again the whole country has served as guinea pigs while the FDA engaged in unconscionable stalling and pretended action.

The FDA's inaction seems even more unjustifiable when you consider that a ban on Red 2 would not even mean an end to red food coloring. For, under pressure from the FDA during the crisis, much of the food industry, including such giants as General Foods and General Mills, switched to another red food dye, Red 40, made and patented by Allied Chemical Corporation. In fact, in September 1971, after the FDA's proposed ban, the production of Red 2 dropped sharply—but it picked up again in February as it became clear that the FDA was not going to take action, and by March

1972 it was back to normal levels. (If given a choice, industry prefers Red 2 because, it says, Red 40 does not give as rich a hue and is slightly more expensive.) It is uncertain whether Red 40 is totally safe, but it does have the advantage of having been put through reproductive tests and is one of only two food colors ever given a permanent safety listing by FDA.

FDA authorities recognize openly, if facetiously, that Red 2 can kill fetuses. When asked about the dangers of Red 2 by a reporter, a top FDA official joked, "People pay good money for that." When the reporter looked quizzical, the official explained, "The pill." That some women may not want forced abortion at random apparently escapes bureaucratic sensibilities. Moreover, as Dr. Meselson has pointed out, early fetal deaths may only be a sign of "an interference with some rather general biological process, such that other ill effects may be discovered as research continues."

That Red 2 is of questionable safety is undeniable: it has been proved a teratogen in rats and chicks, and its cancer potential is suspected. By no stretch of the Congressional guideline on additives—that industry is obligated to prove them safe before undertaking widespread distribution—is Red 2 safe enough to eat. In knuckling under to industry and ignoring and corrupting its scientific and legal responsibilities, the FDA has once again surrendered its protective role. And for what? For an unneeded, frivolous food dye that we can live without.

Cancer in Hot Dogs, Ham, Bacon, Salami, Corned Beef, Bologna, Lox, Etc.?

Dr. William Lijinsky is a scientist at Oak Ridge National Laboratory in Tennessee. Since 1961 he has been working extensively on the problems of nitrites in foods—along with perhaps five or six other laboratories in the world—and is a leading authority on the subject. As he tells of the rats in his laboratory which he has fed combinations of nitrite (found in all kinds of cured, smoke-flavored meats such as hot dogs, ham, bacon, pastrami, corned beef, bologna, nearly all luncheon meats, lox and other smoked fish) and amines (present in wines, tea, fish, in most drugs, including tranquilizers and analgesics, and even in cigarette smoke), he frowns, then talks excitedly. He has found malignant tumors in 100 percent of the test animals within six months, and he thinks they all will be dead in the next three months. "Unheard of," he says in his clipped British accent. "You'd usually expect to find 50 percent at the most. And the cancers are all over the place —in the brain, lung, pancreas, stomach, liver, adrenals, intestines. We open up the animals and they are a bloody mess. I wish you could see the animals—I just wish you could see them."

Is he still eating foods containing nitrites?

"Oh, good heavens, no. In my opinion nitrites constitute our worst cancer problem. I don't touch any of that stuff when I know nitrite has been added."[1]

Dr. Lijinsky isn't eating nitrited food, but most of the rest of us are. Sodium nitrite, as a white granular chemical that is easily mistaken for salt or sugar, or in liquid form, has for the last forty years been used in curing meats. For about ten years it has also been added to some smoked fish. (Sodium and potassium nitrate [saltpeter] have been used for curing meats since ancient Roman days, and still are, to a minor extent in combination with nitrite. Nitrate, also found in water and certain vegetables, notably spinach, beets and broccoli, is of itself not very toxic, but when exposed to bacteriological action in the body or in the environment it is converted to nitrite, which is highly toxic. Our main concern, then, is nitrite, and only incidentally nitrate, in that it may be converted to nitrites.)

Under a strange quirk in the law, the use of nitrite and nitrate in *meat* is governed by the U. S. Department of Agriculture; these two chemicals were given a prior sanction under the 1907 Meat Inspection Act and thus were exempt from the 1958 Food Additive Amendments. The USDA allows a maximum of five hundred parts per million of nitrate and two hundred parts per million of nitrite in finished meat. The use of nitrate and nitrite in *fish* does come under the jurisdiction of the FDA, which allows the same "tolerances" in fish that the USDA allows in meat.

About one third of all the federally inspected meat we consume, or five billion pounds yearly, is cured meat to which nitrite has been added. This includes hot dogs, ham, bacon, corned beef and the gamut of luncheon meats. Under new labeling provisions, the presence of nitrite and nitrate is listed on the package.

The reason for the nitrite in meat, the industry has admitted repeatedly, is primarily cosmetic, to make the meat more attractive by fixing the color a bright, stable red which survives cooking. As the FDA's Dr. Wodicka has testified, "The primary reason for the use of nitrites in both meat and fish is to produce a color that is more stable on storage and heating than is the native color of the muscle of the animal tissue." Through an actual chemical process within the meat, the nitrite is broken down into nitrous acid, which then combines with the hemoglobin of the meat, forming a permanent red color. Industry also claims that nitrite is necessary as a preservative to retard the action of bacteria and prevent botulism. But after extensive hearings in 1971, chaired by Representative L. H. Fountain, a subcommittee of the House Committee on Government Operations reported it could not find "persuasive evidence" that this was true except in special cases such as canned ham—although the point of how necessary nitrite is as a preservative is still in dispute.[2]

Unfortunately, but not surprisingly, nitrite shows little selectivity in its "blood-fixing" properties. It reacts with the blood of living humans just as well as it reacts with the blood in the tissue of dead animal carcasses. This fact was suspected as long ago as the 1880s when a large number of sheep died after eating large quantities of nitrate in foliage. The most peculiar aspect was the darkness of their blood; hence the effects of nitrite-nitrate on living beings was at that time called "blood disease." Today it has a much fancier name—methemoglobinemia, or "inactivated hemoglobin." Physiologically, the nitrite, passed on to humans through cured meats, reacts with the blood hemoglobin to produce a pigment, called methemoglobin, which cannot carry oxygen. The ordinary oxygen-carrying capacities of the red blood cells is seriously depressed, and severe poisoning, even death, can result if nitrite levels are high enough to inactivate enough hemoglobin.

Many persons accidentally subjected to excessive amounts of nitrite have been acutely poisoned. In Buffalo, New York, six persons were hospitalized with "cardiovascular collapse" after they ate blood sausage which contained excessive amounts of nitrite, used to maintain the color in the sausage. In New Jersey, two persons died and many others were critically poisoned after eating fish illegally loaded with nitrite. In New Orleans, ten youngsters between the ages of one and a half and five became seriously ill with methemoglobinemia after eating wieners or bologna overnitrited by a local meat-processing firm; one weiner that was obtained later from the plant was found to contain a whopping 6,570 parts per million of nitrite, whereas the federal limitation is 200 parts per million. In Florida, a three-year-old boy died after eating hot dogs with three times greater nitrite concentration than the government allows. And in Washington, D.C., a man died after eating pure nitrite mistakenly marketed as "Spice of Life Meat Tenderizer," which he accidentally sprinkled on food.

Though these accidents resulted from fairly high nitrite doses, higher than legally permitted by the federal government, there is by no means any certainty that even the permitted lower residues of nitrite in food may not be dangerous, especially to infants, children and others who are more susceptible to hemoglobin damage. In fact, a joint FAO/WHO Expert Committee on Food Additives has warned: "Nitrate should on no account be added to baby foods," and "Food for babies should not contain added nitrite." A. J. Lehman of the FDA pointed out, after a dozen children were poisoned by nitrite weiners in the South, that youngsters are high risks. He wrote: "Only a small margin of safety exists between the amount of nitrite that is safe and that which may be dangerous. The margin of safety is even more reduced when the smaller blood volume and the corresponding smaller quantity of hemoglobin in children is taken into account."[3]

He further calculated that only twenty milligrams of sodium nitrite in a quarter of a pound of meat or fish—which is allowable—can inactivate between 1.4 and 5.7 percent of the hemoglobin of average-size adults. This inactivation, it should be pointed out, is temporary, and the blood does eventually return to normal when the intake of nitrite is suspended; but even in ordinary dosages nitrite suppresses the oxygen-carrying capabilities of the blood in all of us to an extent. And, of course, the effects will be greater on people who eat more than a quarter of a pound of meat or fish at one sitting, and on those who have a disease like emphysema or heart trouble or other defects in the oxygen-carrying capabilities of the blood.

Furthermore, nitrite, at levels at which we are eating it, can systematically and significantly lower blood pressure. In fact, nitrite is a case in which a food additive actually doubled as a drug until very recently. Sodium nitrite used to be prescribed for high blood pressure, and according to the medical literature the therapeutic dose was thirty milligrams, which is the amount legally allowable in about a third of a pound of cured meat or fish. A report made by Hazelton Laboratories in 1960 noted: "The current use of nitrite as medicinals centers around the basic action of nitrite in the body; i.e., relaxation of the smooth muscle, especially those of the finer blood vessels. Therefore, a fall in blood pressure is the most characteristic pharmacodynamic effect."

To illustrate: In one study, when large oral doses of sodium nitrite were given to ten normal persons and to twenty-nine persons with hypertension, the drug's effects were noticeable within three to five minutes after ingestion; the blood pressure fell in normal persons as well as in the hypertense, and this effect persisted for one to two hours. In another comprehensive study of cardiovascular response to nitrite, including both normal and hypertensive persons, 51 percent showed a fall in systolic pressure after nitrite medication. In general the fall began five to twenty minutes after ingestion, and it took forty to 120 minutes for the blood pressure to return to original levels.

This use of nitrite points up sharply that food additives often do have physiological effects—and while these effects may be desirable for particular persons with particular diseases, there is no reason that the general consumer, who eats meat and fish with nitrite, should be similarly medicated.

Additionally, more recently, scientists have found that nitrite can induce headaches in certain people who are unusually sensitive to it. And in 1971 a new alarm was sounded when researchers in Israel discovered that sodium nitrite produced seemingly permanent epileptic-like changes in the brains of rats who ate it regularly. The brain abnormalities showed up and persisted even in rats fed low doses of nitrite —not much more than what a heavy eater of cured meat might consume.

But these effects, as hazardous as they may be to certain of us, are not the danger that Dr. Lijinsky was talking about when he said he did not eat nitrited foods. He was speaking of a more widespread and more recently recognized danger from nitrite: cancer. Cancer of a particularly virulent form which emerges only as a result of a peculiar but all too common chain of events. To clarify: when tested in animals, nitrite by itself has not been found to cause cancer. But how often does a person eat meat or fish containing nitrite in isolation? The stomach, at any one time, is a caldron of activity, and scientists believe that it is here that a chemical reaction involving nitrite takes place that can produce a potent cancer-causing substance. Furthermore, the dangerous reaction can also take place in nitrited meats *before* you eat them.

To induce cancer, nitrite must combine with other chemical entities, called amines, which are breakdown products of proteins and are found both naturally in many foods and synthetically in a number of drugs. What happens is that when nitrite and amines of certain types are put together in a mildly acid solution—as in the human stomach—the nitrite forms nitrous acid, which then reacts with the amines to form what are called nitrosamines. These nitrosamines, as a family, have been proved beyond any scientific controversy to be among the most potent cancer-causing agents ever discovered. Scientific opinion is summed up in Dr. Lijinsky's words: "Nitrosamines are among the most potent carcinogens we know and are certainly the most widely acting group of carcinogens."[4] A recent FDA report noted: "Nitrosamines have been described as one of the most formidable and versatile groups of carcinogens yet discovered, and their role as environmental hazards in the etiology of human cancer has caused growing apprehension among experts."

What makes nitrosamines especially interesting to scientists and especially terrifying for the human race is that, unlike other carcinogens, they don't attack one organ only, such as the lung, but they attack all organs seemingly willy-nilly. And also, unlike other cancer-causing chemicals, they have produced cancers in every species of animal that has been tested—in rats, hamsters, mice, guinea pigs, dogs, monkeys. There seems to be no species resistant to their ravages, and it is the wildest kind of wishful thinking to hope that people might be. As Dr. Lijinsky says, "It is most unlikely that man would be the only resistant species."

He, like some other scientists, believes that nitrosamines,

because of their incredible versatility in inciting cancer, may be the key to an explanation of the mass production of cancer in seemingly dissimilar populations. In other words, nitrosamines may be a common factor in cancer that has been haunting us all for years. The theory is that the stomach acts as a "test tube" for a chemical reaction between the nitrite, the amines, and the normal acid of the stomach to form the destructive nitrosamines which are then disseminated throughout the body.

Research on nitrosamines and their sources is exceedingly new. They were first discovered to be carcinogenic by British scientists in 1956; but it was not until 1963 that a German chemist suggested that nitrites might chemically react in the stomach to produce nitrosamines. Since that time research and concern have accelerated tremendously. Needless to say, human cancer from this source has not been conclusively demonstrated. Some, including the FDA, question whether harmful nitrosamines do indeed form in the human stomach from commonly consumed levels of nitrites and amines. But the mounting evidence is so persuasive that it no longer seems doubtful to many scientists.

Here are the pieces of evidence:

1. Nitrosamines have proved to be powerfully carcinogenic. There are many different kinds of nitrosamines, depending on which kind of amine has been used in the chemical reaction; for example, diethylnitrosamine comes from a combination of nitrite and diethylamine. Of about one hundred types of nitrosamines so far tested, nearly all have proved carcinogenic in animals, and some at exceedingly low dosage levels. Particularly striking, some nitrosamines have produced cancer in animals after only a *single* dose. Additionally, nitrosamines that have not affected a mother animal have passed cancer on to her offspring. For example, when nitrosamines were given to rats near the end of pregnancy, the offspring later developed kidney, liver, brain and spinal-cord cancers. In scientific terminology, nitrosamines are transplacental carcinogens.

2. There is no doubt that amines when mixed with nitrites in the medium of gastric juices—both human and animal—can form nitrosamines. Such a reaction has been produced

both in test tubes and in living animals, including humans. One team of experimenters fed cats and rabbits only two hundred parts per million of nitrite—the allowed amounts in finished meats and fish—along with the amine diethylamine, and upon pathological examination it was found that the nitrosamine diethylnitrosamine had formed in the animals' stomachs.[5] These tests are particularly important, as rabbits and cats have stomachs of approximately the same acidity as humans. And the degree of acidity somewhat controls the rate of the formation of nitrosamines: if it's too high, they don't form; if it's too low, the rate is slowed, though it can take place in near-neutral acidity conditions. Though the acidity of the human stomach varies depending on the amount of acid secreted and the stage of food digestion, in general the human stomach seems an ideal place for such a reaction, say scientists. This was confirmed by experiments in Germany in which thirty-one human volunteers were given both sodium nitrate and an amine, and a nitrosamine (not known to be carcinogenic) did form in their stomachs.

3. The most incriminating evidence comes from tests in which animals fed both nitrites and amines regularly over a period of time develop cancers. More damning, the kind of cancers that develop are the same as when the animals are fed the corresponding type of nitrosamine alone, which indicates that the cancer-producing nitrosamine did form in the stomach by chemical reaction, giving rise to the cancers. For example, rats fed nitrite and an amine, methylbenzylamine, in combination, developed esophageal tumors. And more recently, in the experiments Dr. Lijinsky talked so excitedly about, he discovered that 100 percent of rats fed nitrite and an amine found in a certain drug over a period of time developed malignant tumors within six months! Ordinarily even a powerfully carcinogenic substance does not produce cancers until late in the animals' lives, toward the end of two years, and then only in a small percentage of animals. Fifty percent ordinarily is considered sufficient to label a substance potently carcinogenic. But the most shocking aspect of Dr. Lijinsky's tests is that he used exceedingly small amounts of both nitrite and amine—only 250 parts per million, which is comparable to what we humans are taking in now. (Thus

it can no longer be argued that high doses of the nitrites-amines are necessary to trigger the dreaded chemical reaction, as FDA officials have often asserted.)

4. As further proof, some investigators are beginning to turn up human epidemiological evidence of the cancer-causing potency of nitrosamines. In a certain area of South Africa where an extraordinarily high incidence of human esophageal cancer exists, scientists have recently found that the population drinks locally distilled alcoholic drinks which contain high concentrations of nitrosamines. Additionally, from the biological sphere comes evidence that humans may have about the same susceptibility to the ravages of nitros-amines as rats. It has been found that both rat and human livers metabolize a certain nitrosamine at about the same rate (thus again refuting the old argument that you can't tell much about people from studying rats). The conclusion: there seems little question that if nitrite and amines are pres-ent in the stomach at the same time, nitrosamines will form which in all likelihood are potent carcinogens.

Obviously the next crucial question is: Where do we get nitrite, amines and the acid, the three essential ingredients for this noxious reaction and is this combination likely to be common?

Since acid is always present in the human stomach in some degree, it is a *fait accompli* that we constantly carry with us a kind of ever-ready portable test tube with the first in-gredient already poured in.

The amines, likewise, are nearly omnipresent and are diffi-cult to avoid—though it should be noted that only certain amines, technically called "secondary and tertiary" amines, do react to form nitrosamines, and that not all nitrosamines are cancer-producing, though the vast majority are. Neverthe-less, there's a good possibility that we will take in the kinds of amines that do react; they show up in beer, wine, cereals, tea, fish, in cigarette smoke (which could be dissolved in saliva and swallowed) and quite commonly in drugs—both over-the-counter and prescription. Some scientists theorize that these drugs, many of them taken regularly and over a long period, are a perpetual source of amines in the stomach.

Dr. Lijinsky has prepared a list of over one thousand drugs, which are secondary and tertiary amines that could interact in the stomach with nitrites. They are used for all kinds of purposes: as anesthetics, tranquilizers, high-blood-pressure reducers, diuretics, hypoglycemic agents (to lower blood sugar), muscle relaxants, antihistamines, antidiabetic and antiarthritic agents, antidepressants, oral contraceptives, nasal congestants, analgesics. Among the trade names are such familiar ones as Librium, Contac, Ritalin, Streptomycin, Allergen.[6]

The last component of the deadly trilogy is nitrite—the most dispensable and avoidable. Though you might take in small amounts of nitrites from vegetables and sometimes from water, the far greatest source of this chemical is nitrite as a deliberate additive to meat and fish.

So what happens in the stomach when you sit down to drink a bottle of beer and eat a salami sandwich? Or eat ham if you're taking Contac? Or take an amine-containing drug with your breakfast of bacon and eggs (a common time for taking drugs)? Or perhaps even smoke after a meal containing nitrited meats? (Some researchers speculate that since the specific cancer-causing agent in cigarette smoke has never been found, cancer from smoking may indeed not result from smoke alone but from the amines in smoke which interact with nitrites to form nitrosamines.) So, from all the evidence, the prudent assumption must be that when you ingest nitrite and amines at the same time or in close proximity, so that they remain in the stomach together long enough to interact, you run the risk of creating nitrosamines.

And remember, even though the amount of nitrosamine might be exceedingly small, there is no guarantee that it might not be harmful—for a "no effect" level for nitrosamines has never been established in animal studies. A dose of two parts per million—the lowest level tested—has induced cancer in rats. And, as with other carcinogens, there is a cumulative effect, which has been demonstrated in animals. As Dr. Lijinsky commented, "We have evidence that while the amount of carcinogen might not build up, the effect in the animal body does build up. In other words, the more carcinogen you are exposed to, the more cells are damaged and the more likely you are to develop a tumor within your lifetime.

So I feel no amount of a nitrosamine can be ignored." He stresses also that taking in bits of nitrites regularly over a period of time—as the animals did, and as we do when we eat—provides optimum conditions for developing cancer.

What's more, in some instances we can take in the already formed cancer-causing nitrosamines in the meat and fish we buy. In February 1972 the Agriculture Department and the FDA detected nitrosamines in eight samples of processed meat taken from packing plants and retail stores. Nitrosamines at levels of eleven to forty-eight parts per billion were found in dried beef and cured pork, at five parts per billion in ham, and at eighty parts per billion in hot dogs. More alarmingly, four bacon samples—all different brands—that when raw yielded no nitrosamines, revealed up to 106 parts per billion of nitrosamines *after cooking*. The bacon drippings contained twice that amount. In November 1972 the FDA revealed that further experiments had found high levels of a cancer-causing nitrosamine—up to 108 parts per billion—in four other brands of bacon that had been pan-fried, proving that nitrosamines are widespread in cooked bacon. This caused Dr. Jacobson to label bacon "the worst offender and probably the most dangerous food in the supermarket."[7] The FDA also found nitrosamines in smoked chub and salmon at levels up to twenty-six parts per billion.

The amount of nitrosamines present in the products depended on how much nitrite had been originally added to the meat and fish. The higher the concentration of nitrite, the greater the formation of nitrosamines. Apparently, the nitrite reacts chemically with amines in the meat and fish itself to form the deadly nitrosamines, and for some reason the reaction in bacon is greatly accelerated by heat.

Obviously, neither the FDA nor the Department of Agriculture is unaware of the scientific findings concerning nitrites and their addition to food. Scientists have testified at Congressional hearings, have been included on task forces within the agencies; FDA staff members have written papers on all phases of the possible hazards of nitrites and nitrosamines. Yet both the FDA and the USDA are "studying the matter," treating this hazard as they do others, with obtuse resistance to any action.

In 1972 Dale Hattis, a graduate student at the Stanford University School of Medicine, filed a freedom-of-information suit to obtain all the documents, memos, etc., of the FDA through the years pertaining to its approval of nitrite in smoked fish. This information, contained in his report "The FDA and Nitrite," clearly shows that the FDA was pressured by economic interests into acting illegally and irresponsibly in approving use of the chemical nitrite which the agency itself had long classified as a "poisonous or deleterious substance."[8] To quote an FDA information letter of 1948: "We regard [nitrite and nitrate] as poisonous and deleterious substances not required in the manufacture of any food subject to the jurisdiction of the Food, Drug and Cosmetic Act, and, as such, any food subject to the act and containing any quantity of these chemicals would be deemed to be adulterated under the law, regardless of labeling."

According to law, no "poisonous or deleterious substance" can be allowed in food unless it is "*required* in production of the food" or cannot be avoided by "good manufacturing practice." Nevertheless, the FDA in 1960 began granting permission for the industry to use nitrite and nitrate for purely frivolous purposes: to fix a red color in smoked tuna, then in smoked and cured salmon, shad and sable fish. The manufacturers obviously were hard put to present evidence that nitrite was either safe or necessary, as the law required, but the FDA seemed little perturbed by this oversight. One petitioner, seeking sanction for the use of nitrite in fish, did in fact do a good job of pointing out the medical dangers of nitrite, but the FDA approved its use anyway.

In general, the fish industry's argument, incredibly, was that the FDA should approve nitrite because the industry had been using it illegally all along! One petitioner for the use of nitrite passed off the safety requirements with the statement that "no extensive reports of investigations to establish safety are required in view of the long history in common use and the previously accepted safety of these curing agents in the production of meat and fish products within the already established tolerances." (Actually, the nitrite tolerance of two hundred parts per million approved by the FDA and the USDA for meat and fish is without any foundation in safety; it was arbitrarily designated because in 1925 USDA experiments found that that was the *usual* residue of nitrite in cured hams—therefore it seemed a logical figure to use as a tolerance, they reasoned.)

Then the country was struck in 1963 with a botulism scare in which seven persons died from eating smoked whitefish which had been improperly refrigerated or heated during processing. This crisis gave the fish industry another excuse for wheedling the FDA into further approvals of nitrite for perfectly spurious reasons. The industry argued that some plants, especially the small ones, did not have the proper facilities to process the fish at high enough heat or long enough to kill the botulinum toxin. Thus, nitrite should be allowed as a *substitute*—or, as one industry spokesman put it, as "an additional margin of safety against the development of botulism" *in cases where processors were not following the FDA-recommended procedures for killing botulinum.*

At the same time, industry representatives admitted that most processors were able to get along without nitrite if, as an FDA staffer put it, they "complied with basic sanitation principles and procedures consistent with the production of a safe product." And in fact two states, Michigan and Minnesota, along with Canada, prohibit the use of nitrite in smoked fish in lieu of proper time and temperature processing. Nevertheless, the FDA went along with industry's plea for a more convenient and less costly solution to the botulism problem and granted permission in 1969 for the use of one hundred to two hundred parts per million of sodium nitrite in finished smoked whitefish. In Congressional hearings, the FDA's Dr. Wodicka agreed that nitrite would not be needed if fish were heated to the required 180 degrees Fahrenheit for thirty minutes and distributed under adequate refrigeration. But instead of enforcing high standards, the FDA, to accommodate marginal producers, bent the law requiring a ban on such a poisonous substance when it can be avoided by good manufacturing.[9]

Furthermore, since the FDA exercises little control over the fish industry's use of nitrites, we consume dangerous and illegal amounts in fish. In 1969 an FDA survey showed that of six firms sampled, half were producing fish with dangerously high levels of nitrite, far above the prescribed levels. Residues showed up at 2,844, at 1,457, at 911, at 606 and at 560 parts per million. When the smoked salmon from a Florida firm was found to contain as much as 3,046 parts per million of nitrite, the firm's salmon was destroyed and the firm cited for filth and warned about the excessive amounts of nitrite. The FDA took no action whatever against the other two firms that were in violation. Yet, when questioned in testimony before Representative Fountain's subcommittee, Dr. Edwards replied that such levels might be acutely poisonous, even fatal, for small children, detrimental to a woman or her fetus in the last stages of pregnancy, and injurious to a seriously anemic person.

According to the hearing transcript:

> MR. FOUNTAIN: Dr. Edwards, could such nitrite contents as 2,884 parts per million, 1,457 parts per million, 3,046 parts per million, be acutely poisonous to small children such as two- or three-year-olds?
>
> DR. EDWARDS: I would think there would be a possibility of that, yes.
>
> MR. FOUNTAIN: Just a possibility?
>
> DR. EDWARDS: Well, again I think it would depend on how rapidly they were taken in and in what form they were taken in. That would have a great deal to do with the degree of poisoning.
>
> MR. FOUNTAIN: Could they even be fatal?
>
> DR. EDWARDS: Conceivably, yes.
>
> MR. FOUNTAIN: Could such high levels be detrimental to a woman in the late stages of pregnancy? Either to the woman or the fetus?
>
> DR. EDWARDS: Again I would have to say yes.
>
> MR. FOUNTAIN: Could they injure a seriously anemic person?
>
> DR. EDWARDS: Yes.[10]

On the question of nitrosamines, the FDA maintains a convenient skepticism despite the mountains of evidence, although officials recognize that some nitrosamines now getting into food are potent carcinogens. After dimethylnitrosamine was detected in samples of smoked sable, salmon and shad, Dr. Wodicka was asked if he agreed that dimethylnitrosamine was a very powerful carcinogen, and he said, "Yes, it is." But he also stated that he was "reasonably certain" that the conversion to nitrosamines will not occur in human stomachs at the allowable nitrite level of two hundred parts per million. When challenged by a questioner at Congressional hearings who pointed out that cats and rabbits given only two hundred parts per million of nitrite and an amine produced diethylnitrosamine, and that nitrosamines had been found in human stomachs given nitrate and an amine, Dr. Wodicka said he was familiar with the work. He conceded, "If you are saying that it [nitrosation] may occur, I couldn't deny that." When pressed further as to why he thought there was no human danger from nitrite-created nitrosamines, Dr. Wodicka gave the explanation that the rate of cancer in the human population wasn't going up, whereas if nitrites were wreaking havoc it would be. Actually, the rate of human cancer *is* increasing —but this is flimsy evidence on which to base a link with any particular carcinogen, though the rise undoubtedly is due in part to the careless use of chemical carcinogens.

Unfortunately, the U. S. Department of Agriculture retains the major control over nitrite—over that which is put into meats. And the USDA has long been known for its open economic bias toward agribusiness. In February 1972, Nader associates Harrison Wellford and Peter Schuck and the Center for Science in the Public Interest first petitioned and later sued the Agriculture Department, asking that nitrite and nitrate be removed from all baby food and from all meats as a color fixative, unless and until industry could prove that nitrite was essential in preventing botulism. The USDA replied that there was "a lack of convincing evidence showing any deleterious effect on humans due to the use of nitrite and nitrate in meat or meat food products within the permitted limits . . ."

As far as baby foods were concerned, the USDA said: "The presence of nitrites as a result of the addition of cured meats is at low levels, which should present no potential health hazard through the conversion of hemoglobin to methemoglobin."

On the problem of nitrosamines, the USDA replied:

> The Department was aware that under certain conditions, nitrites do interact with secondary amines to form nitrosamines

and that some nitrosamines are carcinogenic. However, knowledge in this area was limited and analytical methods available to study the possibility of nitrosamine formation in meat food products containing the *permissible* amounts of sodium nitrate lacked the necessary accuracy and reliability to give conclusive results. There was only limited evidence to indicate that nitrosamines might be formed in meat or meat food products containing sodium nitrite and sodium nitrate within the limits specified by the regulations.

Note that the USDA limited its discussion to nitrosamines that might form in the meat itself, ignoring the potential of nitrosamines forming in the stomach after the meat had been ingested.

The department's big argument is that nitrite is essential to prevent botulism in cured meats, though USDA regulations approve nitrite for *fixing color only* and not as a preservative or for controlling botulism. It is only since consumer advocates have begun to challenge the unnecessary use of nitrite as a cosmetic color-fixer that both industry and government are coming forth with the argument that it was being used all along to control botulism. There is a legal snag in this. Such a use of the chemical in meats would be a new one, which, by law, requires approval by the FDA. Since this approval has never been obtained, the USDA at present is violating the law in permitting the use of nitrite for botulism control, as Harrison Wellford has argued. To make such use legal, Agriculture would first have to propose it to the FDA and then submit to a public hearing to prove the necessity of such use. This course of action would bring the issue properly into the public arena, something the USDA had steadfastly refused to do; instead, it has simply proclaimed illegally *de facto* that nitrite is essential in preventing botulism.

Certainly some of the staunchest opponents of nitrite admit that the question of whether a ban on nitrite could expose us to some food-poisoning hazards is still unsettled. But there is evidence that it need not. Though industry and government officials insist that nitrite is essential because it can prevent botulism—which apparently it can do, especially in ground meats such as hot dogs and other sausages and in smoked fish—they fail to point out that it is not the *only* method of preventing botulism in such products. Though it works, it may not be essential. There are other methods of preventing botulism, as the fish industry well recognizes. In fact, aren't meat producers saying in effect that they must have nitrite to compensate for poor handling and sanitary conditions? Since meat has to be packaged under certain conditions to allow the botulinum spores to grow, it seems that the danger from this source could be largely reduced by proper handling and packaging. Another solution is extreme high or low temperature, which prevents botulism (for example, thoroughly cooking bacon, which is almost always done, eliminates any botulism danger); therefore a substitute for nitrites is sterilization and freezing or refrigeration. Some shelf-stable hams need no refrigeration because they have been heat-sterilized, killing all bacteria and potential botulinum spores, yet they still contain nitrite for purely cosmetic purposes. The industry today manufactures without nitrite many bratwurst and breakfast sausages which, one might think, using the industry's argument, lend themselves to botulism danger. The reason these particular sausages do not contain nitrite is that it is not needed as a flavor or color ingredient. Botulism has never been a problem in commercial sausages because refrigeration prevents it.

It also seems perfectly plausible that an ingenious industry —one that can give us such things as Squiggly Whip—can come up with a substitute safe chemical process other than sterilization or freezing to accomplish curing and prevent botulism. Such substitutes, which need more testing, have already been suggested. However, instead of going full speed ahead, the industry-USDA-FDA coalition seems firm in its position that there is no substitute for nitrite, now or ever. Their posture is misdirected at defending nitrites, devising ways to keep it in food rather than ways to get it out. As a stinging report on the FDA's and USDA's handling of nitrite and nitrate, put out by Congressman Fountain's Intergovernmental Relations Subcommittee in August 1972, noted, "Concern over botulism is deterring effective regulatory action."

Said the Congressional report:

> The committee believes industry in the United States has the
> scientific and technological capability for devising adequate proc-
> essing, packaging and sanitation practices to protect the public
> against botulism without the need for questionable chemical
> preservatives. The committee believes too, that until such al-
> ternative methods are devised, both FDA and USDA are capable
> of developing an effective interim program to limit the use of
> nitrite and nitrate to the minimum levels necessary for color
> fixation in those products which do not have the capacity to
> cause botulism.

The FDA and the USDA's concern about botulism, the report said, "does not justify their failure to take protective
action against the excessive or unnecessary use of nitrites and
nitrates." The report, based on extensive hearings before the
subcommittee, also denounced the FDA's lack of prosecution
of nitrite violators, the agency's haphazard surveillance program, and its illegal and scientifically unsupportable allowances of high levels of nitrite residues. The subcommittee
pointed out that a study in England twenty-five years ago
concluded that only ten parts per million of nitrite are needed
for color fixation, yet the FDA and the USDA allow twenty
times that amount.[11] Dr. Lijinsky concurs that nitrite allowances in meat and fish should be cut to the barest min-
imums to reduce the production of nitrosamines both in the
stomach and in the product itself.

Proof that we can live without nitrites is shown by the fact
that nitrite-free frozen meats (hot dogs, salami, bologna)
have been made and sold since 1966 by the Maple Crest
Sausage Company, Rochester, New York, under the Shiloh
Farms label, and distributed nationwide, mainly to health
food stores. A few other local establishments are now also
producing nitrite-free cured meats. No case of botulism has
ever been reported from such cured meats.

At the very least, nitrites and nitrates should be immediately banned from baby foods. It is simply inexcusable that
infants be dosed with nitrite, despite the USDA's protestations that it does not pose a threat. That nitrite does get into
prepared baby foods has been extensively documented. A

1969 USDA survey showed that nitrite was found in baby food up to 28 parts per million and nitrate up to 175 parts per million: in vegetable and ham with bacon (20), Gerber Junior cereal, egg and bacon (18), Gerber Junior vegetable and beef (26), strained cream-of-chicken soup (27), Gerber split peas, vegetable and ham (15), to cite a few examples.

Though some baby-food makers have stopped adding nitrite to their products, it is present in cured meats, such as bacon and ham used in the baby foods. If necessary, baby food could contain uncured meats, since the foods are sterilized, thus eliminating the danger of botulism.* It is well known that babies are highly vulnerable to the hemoglobin disturbances produced by nitrite—and also to the ravages of cancer-causing substances.

The FDA and the USDA should swiftly curtail the use of nitrite to the absolute minimum and investigate thoroughly the true necessity of its use as a preservative. In the meantime, those who want to protect themselves should do as Dr. Lijinsky has done, and simply stop eating all such products known to contain nitrite.

Confronted with the same evidence as our government, Norway banned nitrite and nitrate in hot dogs, effective January 1973, and told manufacturers that bans may be forthcoming on other cured products in which nitrite is not essential to prevent botulism. The actions finally taken by our governmental protectors are in sharp contrast to Norway's: The Agriculture Department, in response to the Nader group's lawsuit, decreed that henceforth, as had been legally required but ignored, nitrite and nitrate would have to be labeled on all meat packages as of February 1973; typically, however, when February rolled around, Agriculture granted a six-month extension for enforcement of the labeling requirements. Yet in November 1972 the FDA had taken official recognition of the hazard in the *Federal Register*: "Recent information indicates that under some conditions nitrites may react with secondary and tertiary amines, which occur naturally in many foods, to form nitrosamines in food and

---

* According to Michael F. Jacobson in his report "How Sodium Nitrite Can Affect Your Health," Gerber and Beech-Nut have stopped using nitrite-cured meat in their baby foods.

perhaps that nitrites may also react with such amines in man's gastrointestinal tract to form nitrosamines. Certain nitrosamines have been found to be carcinogenic in test animals and are therefore potentially hazardous to human health." After stating this formidable danger, the FDA then went on to outline its proposed remedies: to ban sodium nitrate, but *not* sodium nitrite, in smoked cured sablefish, salmon, shad and cod roe, to ban sodium nitrite in smoked tuna fish (because its only function is to fix color), and to ban nitrites in pet foods!

The Long, Hard Fight Against Drugs in Meat

In hearings in December 1971, Congressman L. H. Fountain made the interesting observation that the FDA was using the same argument for retaining the cancer-causing hormone diethylstilbestrol (DES) in feedstuff for beef cattle as the industry had made for keeping the drug in chickens back in 1959. In essence, the FDA, once strongly opposed to the use of DES for chickens, now in 1971 had reversed itself, had adopted industry's position and was in favor of keeping it for beef, despite similar evidence that it produced cancer. Perhaps the case of DES illustrates better than any other how the FDA's protective stance on food additives has gravely deteriorated—and particularly how vulnerable we are to a pharmacopeia of drugs that reach us through the meat supply.

When in 1938 a British biochemist, Sir Charles Dodds, accidentally discovered a synthetic substance that behaved like natural sex hormones, he surely had no idea that he was about to become a hero to our meat and poultry producers. At first DES, as the synthetic was called, was used to replenish estrogen-deficient women, and it went on to have other uses, one of them as a drug against certain forms of cancer. But by the 1940s an ingenious poultry industry had discovered that DES had marvelous side effects: when implanted in the necks of live turkeys and chickens it made them grow faster and fatter on less feed. In 1947, permission to implant

DES tablets in poultry was granted by the FDA under a new drug application. Soon afterward, scientists at Iowa State University reported that a mere "thimbleful of DES in a ton of feed makes cattle gain 15 per cent faster at a 10 per cent savings in feed per pound of grain." And in 1954 the FDA gave cattle growers permission also to use DES.

Even at that time there was no doubt that DES caused cancer when fed to animals, for studies in the 1940s had shown this, and it was suspected also of causing cancer in humans. However, authorities at the time believed that no residues of DES remained in animal tissue intended for human consumption. But by 1955 a new method for detecting residues in poultry found that DES was present at twenty to fifty parts per billion in the livers, skin fat and kidneys of chickens. Following these disclosures the FDA revoked the use of DES in chickens in 1959 as "unsafe" and not allowed under the newly enacted Delaney amendment.

Quite naturally, poultry raisers and chemical companies raised their voices. One chemical company went to federal court to contest the FDA order. In an appeal order handed down in 1966, the judge, in a well-reasoned, scientifically impeccable decision, laid out the reasons for the necessity of the ban. The chemical company argued that the residues of DES in the livers of caponettes treated with DES implants were so minuscule as not to pose a health hazard. But the judge declared:

> The record . . . shows that it may take many years, as much as the greater part of a lifespan, for a carcinogen to produce a detectable cancer, and that the quantity of DES which is required to cause a cancer is presently unknown. All that is positively known is that there is a definite connection between DES and cancer. Furthermore, it was shown that prolonged exposure to even small amounts of carcinogenic substances is more dangerous than short-term exposure to the same or even larger quantities. The Commissioner's finding that DES residues in caponettes are unsafe was based on the testimony of a number of prominent physicians [who] testified that based upon clinical experience and to the extent practicable, no quantity of DES, regardless of amount, should be added to the diet.

The chemical company also argued that estrogen is produced naturally in the body and in many natural foods which could expose humans to greater levels of estrogen than the residues in chickens. The judge's response: "The existence of natural estrogen in foodstuffs does not warrant the intake of DES by a deliberate means of exposure through the implantation of such a drug in a chicken so as to make it tastier and to save feed costs. If estrogens are contained naturally in certain items of diet, there is no justification for adding more by an artificial method."[1] And so, with the full support of the FDA, the courts and cancer experts, including Dr. Roy Hertz at the National Cancer Institute, DES was banned in chickens.

But the use of DES in feed for cattle—and also for sheep —was continued undisturbed, on the assumption that residues were not present in edible tissues. The justification was that ten milligrams of DES fed to beef cattle daily, the permissible amount prescribed by the FDA, would be eliminated from the animal if the DES was withdrawn from the feed forty-eight hours prior to slaughter. Under a 1962 amendment to the Delaney clause, cancer-causing substances could be added to animal feed if "no residue of the additive will be found (by methods prescribed or approved by the Secretary . . .) in any edible portion of such animal after slaughter or in any food yielded by or derived from the living animal." Under the law, manufacturers of the chemical were also obliged to supply a "practicable" test method for detecting residues so that the government could take prompt action if meat was getting contaminated.

However, the government conveniently overlooked that last requirement for DES, as well as for other drugs used in animal feed; and the FDA had no way of knowing whether DES residues were occurring in meat eaten by humans, or whether cattle raisers were complying with the maximum dosage and withdrawal times. In fact, a look at some advertising and industry brochures would have told FDA they were not. Such material actually encouraged cattle growers to exceed the legal limits of DES for "faster gains and greater feed savings."

It was not until 1965—eleven years after DES was approved for use in cattle feed—that the U. S. Department of Agriculture started a small surveillance program to try to determine whether illegal residues of drugs were showing up in livestock and poultry for human consumption. (There is split authority concerning drugs in animal feeds: the FDA approves use of the drug, and the Agriculture Department is entrusted with detecting residues.) By that time, the USDA had come up with a test which could detect DES residues in meat at ten parts per billion and over (though cancer had been caused in animals at only 6.5 parts per billion).

The number of animals sampled at time of slaughter was so small as to be statistically unreliable. Nevertheless, from the inception of the surveillance program, USDA scientists were picking up residues of DES in beef, primarily in beef livers, where the chemical was detoxified. In 1965, out of 558 samples, they found 3 percent contaminated with DES; in 1966, out of 1,023 samples, one percent; in 1967, out of 469 samples, 2.7 percent; in 1968, out of a similar sample size, they found .7 percent; and in 1969, .6 percent contaminated.

Now the FDA did know that these clearly illegal residues were occurring—but virtually nobody else did, until Senator William Proxmire, tipped off by a veterinarian who used to work at the USDA, requested confirmation from the Agriculture Department. The information was then obtained by two Associated Press reporters, and in the summer of 1970 there were headlines throughout the country: "Residues of Cancer-Causing Hormone Found in Beef Supply." Though the percentages were small, anyone could see, as the reporters pointed out, that since about 30 million of the 40 million cattle slaughtered every year were fed DES, hundreds of thousands of beef livers were probably DES-contaminated. Did that mean that consumers were eating DES? "I doubt you could escape that conclusion," said Dr. Joseph Stein, director of the USDA's slaughter inspection division. Also, nobody knew how much DES was getting into beef that the tests were too insensitive to pick up.

Primary control of drugs in animal feed rests with Dr. C. D. Van Houweling, director of the FDA's Bureau of Veterinary Medicine. Van Houweling had known of the residues and bluntly said he had "told the Commissioner—if our objective is never to find any positive samples we might as well withdraw the use of DES right now." It was Van Houweling's view that there would never be a time when the FDA could guarantee DES-free beef, and furthermore he was of the school of thought that held there was no danger in low levels of carcinogens in foods, the Delaney clause notwithstanding. "There are some real advantages to using it [DES] and I don't think at these very low levels there is any danger," he said. In response to one man's claim that he got cancer of the colon from eating DES-contaminated beef, Van Houweling replied, "I don't think there is any more chance he has cancer from eating DES than I can jump from here to the moon."[2] In Van Houweling's view, the residues were caused by cattlemen's disobeying the rules on the withdrawal times prior to slaughter, but he felt they could be "educated" not to violate the law. He expressed his faith in cattle producers by actually *doubling* the amount of DES allowed in feed—from ten milligrams to twenty milligrams daily.

Remember that DES was a *known and recognized carcinogen*; it had produced cancer in several species of animals. And under the law, if residues were appearing, the FDA was obligated to ban the use of DES. But the FDA did nothing. It was to take the full force of several Congressional hearings, a lawsuit by consumer groups and threatened federal legislation to force the FDA to budge.

In March 1971, Congressman Fountain's Intergovernmental Relations Subcommittee decided it was time to try to get some answers from the FDA and the USDA about food additives in general, and medicated animal feed and DES in particular. Why was the USDA taking so few samples for DES? Was it afraid of what it might find? (In the wake of the hubbub over DES, the USDA drastically cut down the number of cattle surveyed for DES to a mere 192 out of 31 million DES-fed cattle slaughtered in 1970.) Why was the residue test so crude? Why were residues occurring? And what did the governmental bodies responsible intend to do about it?

Anticipating the criticism, the FDA and the USDA pulled out their counterplans: officials announced they had perfected a new test which detected DES at levels of only two parts per billion in meat; they planned to sample three thousand sheep and three thousand cattle in the coming year, and they had instituted a new voluntary control program in which cattle raisers had to produce a certificate saying they had not violated the withdrawal regulations. Commissioner Edwards stated he could tell the American public with "reasonable certainty" that there was no danger of cancer from DES, that residues would not occur in the future. But he assured congressmen that if they did he would ban DES (albeit reluctantly, he let it be known), as the Delaney clause required.

Throughout, however, FDA and USDA officials insisted that, despite residues, there was no real evidence that DES caused cancer in humans—only in animals. They bore down hard on the fact that DES had been used in humans for many years without causing cancer, or so they thought. At one point a congressman asked the FDA's Dr. Wodicka, "If you found no evidence of carcinogenicity, in testing animals, but you did find it in human tests, you would certainly take action, would you not?"

Dr. Wodicka: "Yes."

Then the second, more tumultuous act of the DES drama opened.

In its issue of April 22, 1971, *The New England Journal of Medicine* published a bombshell of a study that confirmed the cancer-causing potency of DES in *humans*. It was a remarkable piece of medical detective work that involved tracing an event back some twenty years, to a time in the early fifties when the hormone DES was given to pregnant women, ostensibly to prevent their miscarrying. The mystery began in 1967 when a teen-age girl walked into Boston's Massachusetts General Hospital, complaining of bleeding between her menstrual periods. Dr. Howard Ulfelder, a gynecologist at the hospital, discovered that the girl had one of the rarest of cancers—an adenocarcinoma of the vagina; moreover, this type of cancer is almost unheard of in women under thirty; thus the doctor thought he had merely encountered a freak case. However, within the next two years six other women, all under age twenty-five, showed up at the hospital with the same kind of cancer. Doctors, intrigued by the "coincidence," began a search for the cause. They considered nutrition, genetics, familial similarities, etc., but nothing linked up. Then,

hrough a study of medical records and the recollections of
he girls' mothers, they found the missing factor: the mothers
uring their pregnancies had been given the hormonal drug
ynthetic DES to prevent possible miscarriage. The connec-
ion was so striking as to be unmistakable: it was clear to the
octors that the drug had affected the developing fetus in
uch manner as to give rise to the rare cancer some fifteen to
wenty years later.

The prestigious *New England Journal of Medicine* called
he association a "stunning observation," and did not fail to
ote the implications of the findings for an entire population
hat was exposed to DES through eating DES-fed beef cattle.
:ditorialized the *Journal*: "By avoidance of the prescription
f stilbestrol to pregnant women, this unusual cancer may
e prevented in the future. But more worrisome is the
tilbestrol residue in meat. Of 40,000,000 cattle slaughtered
a this country each year, 30,000,000 have been fed stilbestrol
o increase their weight . . ." The *Journal* went on to point
ut that the surveillance program of the Agriculture Depart-
nent was finding only infrequent residues of DES (an irony
o be appreciated later) but that the danger could not be
ppraised, since even the new tests were not sensitive enough
o reveal *all* residues of DES, and it was well known that "the
etus is so much more vulnerable to minute doses of a car-
inogen."

Alerted by the *Journal*'s report, other doctors further es-
iblished the connection between DES and the rare vaginal
ancer. By the end of the year, sixty-two young women with
he rare cancer, all of them under the age of twenty-four, had
een found, and in nearly all cases it could be conclusively
stablished that their mothers had been prescribed DES dur-
ig pregnancy. Some of the young girls died; others under-
ent serious surgery, removal of the vagina, to save their lives.
'hysicians in New York immediately put out a warning
gainst the continued use of DES in pregnancy, and four
nonths later the FDA finally followed suit and warned that
)ES was "contraindicated" during pregnancy.

The question remaining, however, was: Were we neverthe-
·ss still dosing millions of Americans with harmful amounts
f DES through meat, perhaps condemning not only our-

selves but countless unborn to the ravages of a particularly virulent form of cancer? Weren't we, in effect, medicating an entire population, including pregnant women, with a hormone that had just been declared too dangerous to be used as a drug, but was still "safe" when taken into the body from eating contaminated beef?

Of course not, said the FDA and the USDA. Everything was under control. The new surveillance program was in effect, and they could announce with assurance that, at least as garnered through their new test methods, no DES residues were showing up in red meat. And these pronouncements set the stage for still another shocking revelation concerning DES in 1971.

At the time when the new information about DES causing vaginal cancer was coming to light, the Agriculture Department, at Congressional request, was sending monthly reports to a number of congressmen, including Representative Fountain and Senator Proxmire, informing them that no residues of DES were being found in beef and sheep under the new monitoring system. High officials in Agriculture gave the press the same information. Said Assistant Agriculture Secretary Richard Lyng in an interview, "All tissue samples collected and analyzed in the objective phase of the DES program have been reported negative for DES residues during the period January 1 through June 10, 1971." As late as August 31, Lyng confidently told Senator Proxmire the same thing.

However, David Hawkins, attorney for the National Resources Defense Council, Inc., who had been concerned about DES, began getting inside information from the USDA that the public pronouncements did not match what scientists at Agriculture's laboratories in Beltsville, Maryland, were finding. They were indeed discovering alarming DES residues— at about six times the incidence of the previous two years. In October, Hawkins sent off a letter to Secretary of Agriculture Clifford M. Hardin, charging that the USDA was misleading the press, Congress and the American people by its persistent false statements and concealment of the fact that DES was being found in beef and sheep. Hawkins called this "final proof that DES simply cannot be used safely as an animal drug," and asked the USDA to join in urging the FDA to ban the drug.

Within four days the Agriculture Department issued a press release confirming that DES had been found in fourteen samples of beef and sheep livers during the period in question. Secretary Lyng, obviously embarrassed, admitted that the USDA had made an "inexcusable error" in misinforming the public and vowed to Senator Proxmire that he was "conducting a thorough investigation" and would "pinpoint the responsibility for this gross malpractice." (Later it was reported that officials in the laboratories had not been reporting the residue findings to their superiors in Washington, because of a "technical problem.")

With the new disclosures, pressure to have DES banned from livestock use increased. And the fight went into the public domain. In October, the National Resources Defense Council, joined by several other consumer organizations, filed suit in U.S. district court to force the FDA to ban DES in animal feed; and in November, Senator Proxmire introduced legislation to require an end to feed use of DES.

When confronted with the new evidence linking DES to cancer in humans, the FDA refused to "believe" that such evidence proved a cause–effect relationship (Dr. Edwards termed the evidence of vaginal cancer in young women from DES merely a "statistical association"), and insisted that in any event the DES residues in meat were infinitesimal and were under control. Dr. Edwards repeated that he could with "reasonable certainty" assure the public that it was in no danger of cancer from DES, announcing that the FDA had adopted tighter regulations which virtually made DES residues impossible. The FDA had upped the withdrawal time from forty-eight hours to a full week (effective January 8, 1972) prior to slaughter, and the USDA had instituted mandatory certification by livestock growers that DES had been withdrawn from feed no later than the one-week period.

Many at the time contested the validity of the new regulations, insisting that they would not work and that residues would still appear. Cancer specialist Dr. Roy Hertz, now a senior physician at Rockefeller University in New York City, called the new regulations a "foolhardy undertaking" and said that nothing short of a famine justified the continued use of DES.

In staunchly supporting the continuance of DES, the FDA became embroiled in a scientific controversy with some of its own scientists and those at the National Cancer Institute, and in the midst of it the commissioner wrote letters to the newspapers and gave speeches assailing what he apparently took to be hysteria over the DES question. In November, Dr. Edwards gave a speech before the National Academy of Sciences in which he praised industry and attacked consumer advocates Ralph Nader and Morton Mintz, an investigative journalist for the *Washington Post*. Said Edwards, "The voice of responsible science must rise above the din. If not, the 'special interests,' the zealots, and the extremists will drown out the voice of reason when it is most needed." He said the controversy over DES was a "case in point."

The FDA was lined up against the leading experts in the world on cancer causation. In defending its position and in its flailing efforts to keep DES in livestock feed, it sank further into absurdity. Along with the USDA, it even resorted to trying to win public support by frightening already price-conscious consumers about an increase in the price of beef should DES be banned. Commissioner Edwards stressed the great economic value of DES, stating that "a 500-pound animal will reach a marketable weight of 1,050 pounds using 511 pounds less feed and 31 days sooner when fed DES-containing feed." And USDA economists estimated that banning of DES would raise the cost of beef three and a half cents a pound, which, when multiplied by the per-capita yearly beef consumption of 110 pounds, would mean an added annual cost of $3.85 a person, or $800 million nationally. That these figures were employed as an argument for keeping DES was inexcusable. The implication was, can a little risk of cancer really be worth an $800 million rise in our meat bill? Exasperated, one congressman at one point during hearings told FDA officials to leave economics to the Agriculture Department, where it belonged, and concentrate on the health issue, which was FDA's mission.

Senator Proxmire called "cheap beef or lamb" that brought with it the threat of cancer "a very bad bargain indeed."[3] He further pointed out that the United States had not broken any speed records compared with other countries in eliminating the DES hazard. Twenty-one countries, some as far back as 1959, had forbidden the use of growth-promoting hormones, including DES, as a cattle-fattening agent, because of the health danger. Among them were Austria, Belgium, Denmark, France, Greece, Italy, Sweden, Switzerland—and Argentina and Australia, the principal beef producers of the world besides the United States. A little later, congressmen were shocked to learn Italy and Sweden had even banned the importation of American beef that had been fed DES. The fiction that Americans eat the safest food in the world was quickly going down the drain.

In November and December of 1971, Congressman Fountain, the diligent overseer of FDA activities, recalled the FDA and USDA officials for other inquiries about their laxity in protecting the public from DES. After listening to the FDA's arguments about why the use of DES was perfectly safe, Congressman Fountain pointed out that its attitude was a far cry from the one taken by an FDA of another era, twelve years earlier, when DES was banned in chickens. At that time government officials, with the full backing of cancer research specialists, thoroughly accepted the view that there is no allowable "threshold" for cancer-causing substances in meat or any other food, and also accepted their responsibility to rid food of such chemicals. Dr. Arthur S. Flemming, then Secretary of Health, Education and Welfare, who approved the ban of DES in chickens, said that once a chemical was proved to cause cancer in test animals "there is no reliable basis on which discretion could be exercised in determining the safe threshold for the established carcinogen."

Contrast that with the FDA's position when the question of DES in beef came up. Dr. Edwards categorically told congressmen, "Our scientists as well as many others outside of FDA, do not accept the no-threshold approach to carcinogenic exposure as it applies to an environmental chemical carcinogen." On this issue of such public-health importance, the FDA, perhaps for the first time, was forced into open conflict with leading cancer authorities—including those at the

National Cancer Institute—who, unlike the FDA, had not changed their minds since 1959.

It was in this context that Congressman Fountain chastised the FDA for presenting arguments that only a few years ago —in the conflict over DES in chickens—it had opposed from industry. This reversal of roles makes it clear why outside action—by Congress and consumer advocates—has become increasingly necessary: to fill a vacuum as FDA continues to shrink from its protective role.

Basic to the DES debate was: Is there any danger of cancer in humans from eating DES at levels that were occurring and could occur in beef and sheep? Residues in 1971 were occurring at about one half of one percent of all cattle slaughtered, or about 150,000 animals yearly, at levels ranging from two to a hundred parts per billion. Many animal studies—in five species—had shown DES to cause cancer and a 1964 study had shown that as little as 6.25 parts per billion of DES fed to mice in their diets induced cancer—way below the levels being detected by government scientists. Furthermore, the government's tests were not sensitive enough to detect DES presence below two parts per billion. Therefore, if there were lower residues (perhaps occurring more frequently in musculature, such as steaks) they were escaping recognition.

As the controversy became rough, the FDA sometimes reached for public support by deliberately warping scientific information and trying to make the whole situation seem ludicrous. For example, Commissioner Edwards in a letter to the Washington Post ridiculed the fact that the two-parts-per-billion test might not be sensitive enough by pointing out that it was the "equivalent of 1½ drops in 25,000 gallons of DES." What he failed to point out is that two parts per billion of DES is two trillion (or two million million) molecules of the hormone in a pound of meat. And as far as cancer specialists know, it may take only a molecule or so at the cellular level to stimulate the runaway growth that is cancer.

It is easier to understand if you think of these few molecules of DES entering, not a body that has been previously free of carcinogenic exposure, but one that through the years has been receiving and "storing" other carcinogens. Thus it is probably not a single infinitesimal dose of a carcinogen that alone and by itself induces cancer, but a cumulative effect, as was amply brought out by cancer researchers during the DES controversy. The "predisposition," if one can call it that, to cancer builds up over a period of time, perhaps from constant exposures to carcinogens along with individual susceptibility, and there may come the day when that extra molecule of a carcinogen may overload the system and cancer begins to grow. Researchers from the National Cancer Institute assured Congressmen that it might be possible for only *one molecule of DES* in the 340 trillion present in a quarter of a pound of beef liver to trigger human cancer, as far as they know.

The FDA and the USDA totally ignored the cumulative effects of carcinogens and tried to confuse and win over Congress by stressing that a person would have to eat ten pounds of DES-contaminated liver every day to receive significantly harmful amounts, and that the chances of this happening were remote. They insisted that the doses that had been given during pregnancy to women whose daughters later developed vaginal cancer were much larger than those found in beef. They also tried to minimize the danger, as chemical companies had done in the 1960s, by stressing that hormones of the same type as DES occurred naturally in the body and in alfalfa and other vegetables, implying that since these were naturally occurring substances there was no danger. Dr. Henry F. Simmons, director of the Bureau of Drugs, said that the amount of hormone produced daily in a woman's body was more than that ever found in *any* residue in meat. Secretary of Agriculture Earl Butz, in talking to cattle growers, took this misleading message nationwide.

All of these arguments were systematically demolished by experts. Regarding the natural-hormone question, Dr. Hertz, who had been instrumental in making the FDA's case for the removal of DES from chickens years before, must have felt a sense of *déjà vu* as he reiterated the views of the scientific community. He noted that it had long been known that it was not necessarily the presence of the natural hormones but the *extra burden* from hormones which was necessary to induce cancer experimentally, apparently by upsetting the body's normal hormonal balance.

How much of the extra dose was necessary to incite cancer was unknown, he said. "But we do know that if we disturb the balance by various physiological experiments, thereby doubling what the body produces itself, we can, under certain experimental conditions, elicit malignant changes in the peripheral tissue. Hence, we have to consider that any impingement on the normal balance of what is produced in the individual herself or himself is a substantial hazard . . ." Furthermore, DES is not natural, but is a man-made synthetic with different properties. Said Dr. Frank Rauscher, director of the National Cancer Institute, "DES is chemically different from natural estrogens and we do not know its molecular mechanism of action."

A parade of scientific witnesses before Fountain's committee consistently contradicted FDA views on the hazard of DES in meat. Dr. Saffiotti of the National Cancer Institute testified that the carcinogenicity of DES was so well established that "it is one of the chapters in a textbook on cancer research." In opposing even the minutest amounts of carcinogens in foods, he said: "I am one of the scientists in the field who have made a strong case for being very prudent and for not allowing any known chemical carcinogens in food use. I think this is a very prudent policy which is being widely accepted by people in cancer research and I would certainly like to restate it now." When asked whether DES should be taken off the market, he replied that if controls could not guarantee that "there is no transfer of the compound to human food under any circumstance," then we "would have to take the next most drastic step and remove it at its source . . ."[4]

Dr. Hertz stated bluntly that in his opinion, since no amount of a known carcinogen could be presumed safe for humans, it was "ill-advised, unfeasible and foolhardy" for the FDA to continue to allow DES in animal feed. "I would say the hazard of any amount [of DES] in any foods to be ingested by any human subjects . . . would constitute a substantial hazard which we would be much better without."

When Congressman Fountain pointed out that since liver was used in prepared baby foods, DES possibly could be a hazard to infants, Dr. Hertz agreed that the younger a person

s, the more susceptible he generally is to carcinogens. As another scientist has observed, one carcinogen caused tumors in
infant mice with a dose only 1/4,000 the size of that required
to induce tumors in adult mice. Then, of course, there was
the problem of fetuses, which, scientists noted, are "exqui-
sitely sensitive" to cancer-causing chemicals. As Dr. Rauscher
said, "It is unfortunately quite possible . . . that men and
women who have been exposed to DES during fetal life will
represent a high risk for cancer. The whole nation is a research
laboratory when it comes to this kind of problem." Dr. Peter
Greenwald, director of the Cancer Control Bureau of the
New York State Department of Health, told congressmen that
most assuredly he would not want his wife to eat beef contaminated with DES were she pregnant.

Finally, certain scientists within the FDA became so embarrassed and concerned about the FDA's lack of action and
its unsupportable stance that they too rebelled. In a series of
internal memos, Dr. M. Adrian Gross of the FDA's Bureau
of Drugs severely criticized the FDA's posture and distortion
of scientific information before Congressional committees. He
called the FDA's position "absurd and indefensible" scientifically, and expressed the strong view that the only way to
control DES was to ban it from feedstuffs. Though the FDA
attempted to suppress these documents of internal dissent,
they did find their way to the Fountain committee.[6]

Through it all, Commissioner Edwards steadfastly maintained that there was "no evidence that these small amounts
of DES will produce cancer," that the "exposure is absolutely
minimal" and that "the risk of cancer, is, for all practical purposes negligible." He asserted that the FDA, after all, had
been doing a good job in keeping DES residues out of 99.5
percent of the beef (although, of course, the Delaney amendment makes no concession for any amounts of cancer-causing
residue). As 1971 drew to a close, the Commissioner assured
Fountain's committee that if the stiffer controls—the length-
ened withdrawal time and stepped-up prosecution of violators—proved unsuccessful, the FDA was "prepared to ban the
use of DES entirely in animal feeds."

It was not long before congressmen were to recall this promise. Ironically, under the "new controls" on livestock growers

(coupled with a more sensitive detection system), more DES than ever began showing up in meat. The controls had proved to be, as Dr. Hertz predicted, "a foolhardy undertaking." In the spring and summer of 1972 the incidence of DES rose to 2.27 percent in beef, or, as Senator Edward M. Kennedy observed, four times the frequency during 1971. It could be estimated, he pointed out, "that more than 660,000 head of cattle probably reach American dinner tables in various forms with a known cancer-causing agent in them."

Still the FDA stood fast, except to announce forthcoming public hearings to consider the evidence against DES. It was, consumer advocates charged, another "stalling device"; for example, one FDA hearing did not take place until fifteen months after it was scheduled. In an interview with *Washington Post* reporter Morton Mintz, Dr. Rauscher, head of the National Cancer Institute, called for an immediate ban on DES, terming it the "prudent" course. Commissioner Edwards was reportedly angered by this public disagreement.

In July 1972, FDA and USDA officials were called up before Senator Kennedy's Subcommittee on Health of the Committee on Labor and Public Welfare, to explain why they did not take Dr. Rauscher's advice. Senator Kennedy let it be known that he thought the FDA was acting irresponsibly and perhaps out of political motives in not wanting to ban DES prior to the upcoming Presidential election in which President Nixon was a candidate. (A ban would not only anger cattlemen but cause a rise in beef prices a week or so prior to the election.) Dr. Edwards' explanation was that DES was not an "imminent hazard" to health and thus could not be banned instantly without a hearing.

Finally the long hassle over DES drew to an end. On August 2, after coming up with "new evidence" that DES was not eliminated from cattle even when it was withdrawn from feed seven days prior to slaughter, the FDA announced that it was ordering an immediate end to the *production* of DES for use in livestock feed. However, the agency continued to allow DES implants in cattle's ears with a 120-day withdrawal period—which would still give rise to residues—and gave cattle raisers until January 1973 to "phase out" their use of DES in feed. In other words, industry was given a five-month grace period in which to exhaust its supplies "as a precaution against disruption of the meat industry and unwarranted public concern," in the commissioner's words.

There was a flurry of protest over the extension and the implant use. Several members of Congress, including Senators Proxmire and Kennedy, introduced bills to get an instant ban on *all* DES use, and the General Accounting Office, the investigating agency for Congress, even ruled that the FDA's "phase-out" action was illegal. But the special bills were blocked in Congress, and were inappropriate anyway, according to Congressman Fountain. In his view why should Congress have to waste time on special bills to force a regulatory agency to take an action already required of it under existing law?

And so, finally beaten by law, the FDA succumbed, and in January 1973 banned DES from animal feed. In April, it also banned DES animal implants, after residues were found. It had taken eight years, several Congressional hearings, a consumer lawsuit and threatened federal legislation to do what, as Representative Fountain pointed out, is the regular, ordinary function of the FDA.

However, the DES case was not yet over. Industry took the matter to court, and in January 1974, the U. S. Circuit Court of Appeals in Washington, D.C. decided that the FDA had violated the law by not giving the users of DES the hearing they were entitled to. Consequently, the FDA's decision was rendered ineffective and cattle raisers were once again free to use DES under FDA's strictures of withdrawal time. In February, 1975, the USDA detected residues in beef by using quite a sensitive method. So, the withdrawal time that DES could be fed to cattle prior to slaughter was stretched from even to fourteen days.

One would think that after the court decision, the FDA would have quickly moved to set up the hearing with industry the court dictated, and dispose of the DES matter once and for all. But it didn't—and hasn't. By April 1975—more than a year after the court's decision—the FDA had scheduled no hearing. And the USDA is still finding residues of DES in beef, despite the new withdrawal times. Why hasn't FDA scheduled a hearing? Critic David Hawkins, who fought the

DES battle for so long, says the FDA, without public an
Congressional pressure is simply ignoring the whole matte
Mainly because it can't decide precisely what methodolog
to use in determining the risk of DES. If it goes into an i
dustry hearing promoting the current method it is using, h
says, DES is bound to be declared illegal. But more impo
tant, using that method, the FDA would be stuck with als
banning the numerous other growth-promoting drugs for an
mals that the agency now allows.

For even if DES were outlawed tomorrow, there are abo
a hundred other drugs given to animals through their fee
many of them as potentially dangerous as DES. For exampl
there are fourteen other synthetic hormones like DES bein
administered through their feed to food-producing animal
Many scientists believe these have the same cancer causin
properties as DES, and even the FDA has classified ten of tl
hormones as proven or potential carcinogens.

One of these synthetic hormones is melangesterol acetat
called MGA, used in heifers to stimulate growth. Accordin
to an internal memo dated September 27, 1972, to Dr. V.
Houweling from the director of the FDA's Bureau of Ve
erinary Medicine, FDA scientists are concerned about tl
drug; and they believe that the forty-eight hour withdraw
time from feed should be extended to thirteen days. Th
note in the memo that the "Upjohn Co. has submitted da
showing that MGA induces tumors in female mice . . ."

Another case is dienestrol diacetate, which, like DES, h
been implicated in vaginal cancer in the daughters of wom
treated with the drug during pregnancy. It is mixed into t
feed of chickens and turkeys, and though residues are like
the government has no test for detecting them. This idio
has been commented on by attorney David Hawkins in Cc
gressional testimony: "Like DES, dienestrol diacetate l
been the subject of an official FDA warning to physicia
contraindicating its use in pregnant women. Yet this sar
FDA continues to permit the unchecked use of this sar
drug in our food supply with no way of telling whether it
reaching our dinner tables."

And if you still believe that you are getting antibiotics o
when they are prescribed by your doctor, consider that,

credibly, about half of the nation's entire antibiotics output, or $50 million worth yearly, goes not directly to humans but into feedstuff for animals intended for human consumption, to make them gain weight faster and keep them disease-free. There is no question that these antibiotics, like DES, are sometimes deposited in animal tissue, and hence are transmitted to our own bodies. In 1969 a USDA study found illegal residues of penicillin in 130 of 470 animals tested—about 28 percent of them! In 1972, illegal residues of streptomycin showed up in the veal of 12 out of 264 calves sampled —at levels from 3.5 to 26.3 parts per million, though the permissible residue is *zero*. Organic arsenic consistently shows up in the livers of chickens and pigs, according to residue surveillances by the USDA.

In the September 27, 1972, memo FDA scientists listed thirty-three drugs which, if misused, can leave illegal residues in cattle tissue; these include oxytetracycline, penicillin, progesterone, sulfachlorpyridazine, testosterone, tetracycline, tylosin, zeranol. They also listed forty-seven drugs, five of them possible carcinogens, which could show up in poultry tissue, twenty-seven in swine tissue and fourteen in sheep, twenty-six which may get into milk, eighteen into eggs, and even two into fish. In nearly all cases the scientists complained that the controls—mainly the withdrawal times of the drugs in animal feed—were inadequate to prevent the residues and assure the safety of the food.

As medical experts have stated, these residues of antibiotics pose serious questions for human health. Some persons who may inadvertently get the drugs, notably penicillin, are allergic to them and can have severe reactions. Antibiotics in the stomach may also upset the gastrointestinal process. But primarily scientists are concerned that, through constant exposure to antibiotics, dangerous bacteria may become resistant to the drugs' effects, thus rendering the drugs useless in treating human disease. Already there are a number of deaths on record in which drugs failed to work against resistant bacteria.

Alarmed over the growing threat, British scientists in the Swann Report in 1970 urged an end to antibiotics which are used by humans, such as penicillin and tetracyclines, in ani-

mal feed. The same year, following the British lead, the FDA
set up an Antibiotics Task Force, which finally, after many de-
lays, issued a similar report in 1972, also urging stern restrictions on drugs in animal feed. At this writing, a year later,
the FDA has taken no action—as consumer advocate Harrison
Wellford points out, "not even to curb the key human disease
fighters like penicillin and streptomycin."

How disheartening the situation is was sharply pointed up
in an exchange during 1971 Congressional hearings between
Representative Fountain and a USDA official. Noting that
the government permits a number of drugs in medicated feed,
which it never then checks for residues, Fountain asked Dr
Victor H. Berry, the USDA's acting deputy director of field
operations, "For instance, Food Additive Regulation 121.205
permits the use of reserpine in medicated feeds for broiler
and replacement chickens. Do you check for reserpine as a
residue . . . ?"

DR. BERRY: No, Mr. Chairman.

FOUNTAIN: Now, Food Additive Regulation 121.207 permits
the use of the drug zoalene in medicated feeds for turkeys. Do
you check for zoalene?

BERRY: Not that specific drug, Mr. Chairman.

MR. FOUNTAIN: How about the food additive ronnel, permitted by Food Additive Regulation 121.209?

DR. BERRY: We have checked for that group, yes, sir.

MR. FOUNTAIN: Or amprolium, covered by Food Additive Regulation 121.210?

DR. BERRY: No, sir, not that specific one.

MR. FOUNTAIN: How about nihydrazone, permitted by Food
Additive Regulation 121.237?

DR. BERRY: No, sir, I would have to say not that specific one
either.

MR. FOUNTAIN: Do you check for furazolidone, which is permitted by Food Additive Regulation 121.255?

DR. BERRY: We have not checked for it . . .

MR. FOUNTAIN: How about testosterone [a suspected carcino-
gen]?

DR. BERRY: No sir. Its uses are very limited—we have put our
resources in testing for other drugs rather than that specific one

When Congressman Fountain asked specifically about a
group of drugs called nitrofurans, USDA official Dr. Harry C

Mussman replied, "Our most recent information regarding detection methods for these residues of nitrofurans in tissues is that at the present time a good technique is not available. There are techniques, but they are not the type that we could use routinely in a regulatory manner."[7]

The logical question then is, Why are these drugs permitted in animal feeds, in direct contradiction to the law which states they cannot be used unless producers provide a practicable method for detecting residues? If it were adhering to the law, the FDA would have to suspend the use of many such drugs until there were good test methods for residues which the government could use in monitoring illegal uses. Furthermore, as the Antibiotics Task Force recommended, the FDA should promptly ban the use of certain antibiotics for animal feeds until and unless it is proved that there is no human danger from their use.

Thus, the long and hard struggle to ban DES resulted in no victory at all. As long as the FDA and the USDA continue to systematically violate the law, DES plus a string of potentially dangerous drugs will continue to plague us as unknown hazards in our food.

Poisoned by Accident: Incidental Additives

Some poisons get into food purely accidentally, and we may never know they are there. A striking example are the poly-chlorinated biphenyls (PCBs). These are synthetic industrial chemicals which were introduced around 1930 as cooling agents in electrical transformers. They are so amazingly stable, heat-resistant and nonflammable that their use spread to inclusion in paints, condensers, oils, adhesives, plasticizers, hydraulic fluids, printing inks, carbonless carbon paper and other products in which they were technically desirable. Manufactured in the United States by Monsanto Chemical Company under the trade name Aroclor, these chamicals have been called "virtually irreplaceable" because of their "unique physical and chemical properties."

The only problem is that in chemical structure PCBs resemble DDT, and, like that insecticide, they are persistent. They hang around the environment a long while, and they are stored in body fat. As an FDA staff paper on PCBs stated, "Like DDT, PCB will enter the organism, be stored in the fat and thus be transferred through the food chain at increasing concentrations." Consequently, because we eat PCB contaminated food, most of us are carrying around small residues of PCBs in our bodies. Because PCBs are virtually indestructible, they are difficult to dispose of once the produc containing them comes into disuse or disintegrates. Like DDT, the PCBs remain behind. They seep into the earth, are carried into rivers, lakes and oceans, and sometimes are carried in the air. They are accidentally ingested by tiny organisms, which are then ingested by larger ones, and the PCBs move up the food chain in increasingly greater concentrations. Some scientists now believe PCBs to be our most widespread environmental contaminant, more prevalent than DDT and much more toxic.

The wide environmental contamination by PCBs came to light in 1966 when a Swedish scientist, Dr. Sören Jensen, identified them in the feathers of birds, in sea eagles and in two hundred fish specimens. (Previously scientists had identified what we now know are PCBs only as "unknown but chlorine-containing compounds.") With publication of Dr. Jensen's findings, other scientists took up the search—most notably Dr. Robert Risebrough of the Institute of Marine Resources at the University of California at Berkeley. Dr. Risebrough found PCBs in sea birds off the coast of California, in fish, and in human milk of nursing mothers in California at an average concentration of 2.6 parts per million.

That humans in general are contaminated with PCBs has been documented. One study by the Environmental Protection Agency (EPA) of 723 people in one South Carolina county showed that 40 percent had PCBs in their tissues at concentrations of twenty-nine parts per billion. Another EPA study of 688 people in Michigan, Florida and Colorado discovered that almost two thirds had PCBs in fatty tissue ranging from a trace to more than two parts per million. Thus it's safe to say, as the government has said, that more than half of us are contaminated by PCBs.

PCB contamination has been recognized in government circles for some time, due to sporadic but unpublicized outbreaks. But it was not until the summer of 1971 that the PCB problem crashed to public attention on the East Coast. At Holly Farms, a leading producer of broilers in Wilkesboro, North Carolina, Dr. Kenneth May, director of research and quality control, discovered that their chickens weren't hatching properly and many were sick and dying. Dr. J. R. Harris of North Carolina State University, who tested some of the chickens, described them:

> The most common symptoms were gasping and incoordination of the head and neck just prior to death. The most severely affected birds emitted a squeaking sound. The mortality ran to 50 percent and the remainder were stunted to the point that they were destroyed. . . . On postmortem examination, most birds had a severe hydropericardium with sufficient fluid in the peritoneal cavity to cause distention. A greenish yellow gelatinous fluid was found under the skin of some of the birds. The livers showed varying degrees of degeneration, with a mottled roughened appearance resembling cirrhosis. The kidneys were swollen with small hemorrhages. Some of the birds had white caseous-like material, resembling a bacterial infection, over the heart and liver.

Dr. Harris also noted that the breeder flocks ordinarily hatched 88 percent; some were now down to 5 percent.[1]

Investigators suspected a toxic substance in the feed, and through tests from the East Coast Terminal Company in Wilmington, N.C., it was isolated: PCBs. The feed contained as much as 148 parts per million of PCB, and analysis determined that it was the toxic factor in the affected chicks; embryos were killed by as little as one part per million or less of the PCBs.

Holly Farms' Dr. May contacted Monsanto Chemical, and in the late afternoon of July 16 Monsanto called the FDA to inform them of the hazard. FDA investigators were dispatched to the East Coast Terminal Company and confirmed that the fish meal was highly contaminated with PCBs. To prevent the fish meal from being infected with salmonella (a food-poisoning bacteria), the company had begun pasteurizing the meal by passing it through a heat-exchange system containing PCBs; the dangerous chemicals had leaked into the fish meal. The company had been attempting to fix the leak since April, and had used about five thousand pounds of PCBs in June alone. Between April 30 and July 16, the company shipped about sixteen thousand tons of presumably contaminated fish meal, which was mixed into approximately 800,000 tons of feed eaten by poultry and hogs in a wide area of the South. The unused fish meal—about one thousand tons

—was recalled, and Holly Farms condemned about 88,000 chickens in the last two weeks of July.

There is no doubt that PCBs got into hundreds of millions of chickens and hogs fed the contaminated feed, and hence to consumers. USDA samples showed that at slaughterhouses across the South, about 70 percent of the broilers which had been given the feed were contaminated with "traces of PCBs" up to 1.4 parts per million. Millions of feed animals marketed during the leak had passed through the slaughterhouses and on to supermarkets directly or to processors for inclusion in canned and frozen foods such as soups and TV dinners.

Eggs too, similarly contaminated, reached the market and were consumed. The FDA immediately put an embargo on about 450,000 eggs known to be contaminated above .5 parts per million, but through a comedy of errors sixty thousand of the tainted eggs were sold in the Washington, D.C., area even though the distributor was less than six blocks from the FDA's laboratory headquarters. And, of course, countless eggs had been distributed and eaten by unknowing consumers before the leak and the contamination were detected.

Government regulators soft-pedaled any danger. At first they withheld announcement of contamination. Then Dr. Yeutter of the USDA said that the public "should not be unduly alarmed." He added, "It is not worthwhile to worry the consumer now about something he ate a few weeks ago. It hasn't affected my consumption of chicken."[2] The FDA hastily declared "guidelines for safety" for the contaminated chickens and eggs—of five parts per million for chickens and .5 parts per million for eggs. (If contamination was under this level, the FDA would not take legal action to remove the product from the market.)

These "safety" guidelines, as a congressman later noted, were "pulled out of a hat" for immediate regulatory purposes. There is no toxicological data showing such a tolerance to be safe for humans. The five-parts-per-million figure is based on *acute* toxicity studies; it certainly doesn't protect consumers against chronic effects or the fact that these chemicals are *cumulative* in the body. Two parts per million today, three more tomorrow and another on Sunday, and your body has already been assaulted by an "unsafe" dose. The guidelines, lacking any scientific validity, were intended only to prevent the wholesale destruction of contaminated chickens and eggs.

There were also discrepancies in the reported degree of contamination, as Harrison Wellford pointed out to congressmen. The USDA insisted that none of the chickens it analyzed from Holly Farms contained more than two parts per million of PCBs, which was well below the "guideline." However, Holly Farms personnel told Wellford that *all* the chickens they had tested were well above the five-parts-per-million "safe" level, and that they had found residues up to forty parts per million in their poultry.[3] If this was true at Holly Farms, it seems certain that it was equally true in other broiler farms using identical feed.

How toxic are PCBs? Studies show that only five parts per billion of PCBs in water kill 72 percent of the shrimp in twenty days. Experiments on rats given PCBs reveal liver damage and reproductive interference, mainly fewer offspring who live a shorter time. When Coho salmon containing fifteen parts per million of PCBs were fed to mink, the mink died within sixty days. (Apparently mink are extraordinarily sensitive to PCB toxicity.) In chickens, besides being immediately toxic—as evidenced by the pathetic deaths of chickens fed PCBs at Holly Farms—PCBs cause birth defects and a syndrome called "chick edema" disease which has been studied since the late 1950s; the baby chicks are born with huge sacs of fluid on their necks, rumps and backs and a mass accumulation of fluid around their hearts. PCBs, like another powerful chemical, called dioxin (a contaminant of some herbicides), also produced gross skeletal defects in baby chicks.

We even have striking human evidence of PCB poisoning. In Japan in 1968 about three hundred people developed a peculiar and severe kind of skin disease; the cause was traced to a shipment of commercial rice oil, used for cooking, which had become contaminated with PCB. The victims used the oil for about a month before symptoms appeared. Nine of the victims were pregnant and two of the babies were born dead, each showed signs of chlorobiphenyl poisoning.

In Senate testimony, ecologist Risebrough calculated that to get the same amount of PCBs as the poisoned Japanese, a person would have to eat forty-five pounds of fish containing the government-allowed five parts per million. But it's well known that many foods, notably fish, exceed that toxicological allowance. For example, Sam Spencer, assistant chief of fisheries of the Alabama Department of Conservation, confirms that fish collected in his state's waters in 1970 showed contamination as high as 365 parts per million—*seventy-two times the FDA limitation.* Catfish, he said, had levels up to 277 parts per million. Though the FDA made these tests, Spencer claims that it tried to hide the results by not reporting them back to state officials until he made a special request. He says, "I feel the consumers of fish or any other food product are entitled to know what they are eating."[4]

To illustrate the possibility of acute PCB poisoning in this country, Risebrough made these additional calculations: the average concentration of PCBs in the flesh of fish sampled in Alabama was ninety parts per million. Therefore, a person would have to eat only *two and a half pounds* of such fish to ingest the level of PCBs that poisoned the Japanese.[5]

The Holly Farms incident was not the first or last time that government officials were confronted with PCB accidents. In 1969, the FDA found PCBs in milk in West Virginia; grazing dairy cattle had become contaminated with PCBs used in a herbicide-spraying operation along highways. In 1970, PCB residues showed up in milk in Ohio, Florida and Georgia. The contamination was traced to PCB-containing sealant in silos which had migrated into silage given to cows. In early 1971, high levels of PCBs showed up in chickens in New York State; it was thought that they came from bakery wrappers somehow ground up in the chickens' feed. One hundred thousand chickens had to be destroyed.

Soon after the Holly Farms incident died down, there were two more outbreaks in August of 1971—one in Minnesota, where fifty thousand turkeys had to be condemned because of PCB contamination, and another in Oklahoma, where contaminated poultry was discovered. The cause was not clearly established in either case. In 1972, PCB residues were found at high levels in large numbers of chickens in Maine and turkeys in California that had already been slaughtered. Again, the source could not be found.

In some instances, food processors themselves spot the contamination in their laboratories. Swift and Company reported the Minnesota incident, and the Campbell Soup Company detected the PCBs in the New York chickens. Such companies say they reject the contaminated poultry regardless of the FDA's lenient attitudes. A Campbell Soup Company spokesman was quoted in the *Washington Post* as saying, "Contaminated substances did not then, have not now and never will get past our laboratories." An unidentified spokesman for a leading food-processing company said he thought the government's guideline for PCBs was excessive in the extreme. He was quoted as contending, "The government may set five parts per million of PCB as an acceptable level, but we set our tolerance at zero."[6]

Apparently some food companies feel strongly about the intrusion of what they consider foreign chemicals, though they may not be so averse to chemicals that have been clearly designated for food use. But the point is, once again the government has erred far beyond economic necessity in setting PCB allowances (as it has done with other incidental additives, including filth, as we shall see), providing a margin of danger that is unacceptable to certain segments of industry. This fact alone should alert the FDA to the idea that the guideline is too high; nevertheless, the agency has proposed to give the unofficial guidelines the status of legal "tolerances," no longer amenable to ready lowering by the FDA should officials be so inclined.

The FDA has proposed the following tolerances in food for what it calls "unavoidable residues of PCBs": milk and dairy products, 2.5 parts per million; poultry, five ppm; eggs, .5 ppm; finished animal feed, .5 ppm; fish (edible portion), five ppm; infant and junior foods, .1 ppm; food-packaging material, five ppm.

The pervasiveness of PCBs in our food is well documented. Between July 1, 1970, and June 30, 1971, the FDA found that of nine thousand samples of food, five hundred were positive for PCBs—about 5.5 percent. They discovered PCBs in fish (315 of 570 samples), cheese (one part per million), milk (sometimes as high as twenty-eight parts per million), eggs, processed potatoes, oysters, crabmeat, cat food, candy (probably migrated from wrappers), meat, poultry and cereals. In 1972, the FDA found that 5.2 percent of the milk sampled was contaminated with PCBs, ranging from traces to 2.8 parts per million.

In July 1971 FDA surveyors doing a routine analysis for pesticide residues turned up PCBs in a shredded-wheat biscuit. The source of PCB contamination turned out to be the cardboard dividers in a cereal box that had been made from recycled paper, including carbonless carbon paper—which contains thirty thousand parts per million of PCBs. The PCBs had "migrated" into the food. A subsequent survey found that PCBs inadvertently were getting into food packaging for all kinds of products, and into the products themselves: crackers, bread crumbs, macaroni, pretzels, chips, breakfast cereals, prepared mixes, dried milk, dessert and pudding mixes, infant dry cereals—75 percent of them were contaminated—cookies, rice, oatmeal, chocolate and cocoa products, grated cheese, dried fruits, frozen fruit juices. All told, about 19 percent of the food sampled was contaminated with low levels of PCBs —an average of .1 part per million. Since much of the contamination occurred when the food came into contact with recycled paper, the FDA proposed prohibiting recycled paper in food packaging. But the industry says that PCBs today have been significantly lowered in recycled paper intended for food packaging.

When we eat this contaminated food, our bodies become reservoirs for a slow build-up of PCBs. It has been shown repeatedly that the larger species accumulate the largest residues as the chemical is carried up the food chain. A recent study of lake trout dramatically showed that big fish who eat small ones contaminated with PCBs soon end up the most contaminated of all. Smaller, younger fish contained only one or two parts per million of PCBs. Those nine years and older had a lifetime storage of 30.4 parts per million in their flesh.

Cooperative experiments conducted by the USDA and the FDA also show that feeding low levels of PCBs to chickens causes a slow buildup of PCBs in their eggs. If the chickens are given ten parts per million in their feed, within two weeks their eggs also contain close to ten parts per million. At these low levels of PCBs there is no apparent observable disease in the adult chickens—no suspicion that either the chickens or their eggs are contaminated. Thus such eggs could be and undoubtedly are passed on to consumers in the marketplace without any warning signs of contamination. Furthermore, once the PCBs build up in hens' tissues, it takes at least eight or nine weeks after the PCB-contaminated food is withdrawn for the PCBs to disappear from their eggs.

Obviously, PCBs are a widespread environmental problem. No one intended them or wants them as food additives, including the FDA. But unlike other chemical additives, PCBs fall under no one jurisdiction. Consumer advocate Harrison Wellford noted that government control of PCBs is hopelessly split: the USDA is charged with looking for PCBs in meat and poultry, the FDA with looking for them in whole eggs; the USDA in cracked eggs, the FDA in fish meal, the EPA in pesticides. Authority is confused and often lacking. A legal loophole allows these environmental chemicals to slip through —and into our food; it should be closed. Senator Warren G. Magnuson has introduced an amendment to the Federal Hazardous Substances Act which would require that all new substances—"some of which may be a danger to human health or the environment"—be tested for safety and that a constant surveillance system for detecting their presence in food and the environment be established.

In the meantime, we're stuck with PCBs, which are non-biodegradable. Monsanto announced in 1971 that it would no longer sell PCBs to any food or feed company, which was a responsible gesture, though it seems impossible that Monsanto can retain tight control over PCBs once they are shipped out of its plant. In March 1972, at the same time that the FDA proposed PCB tolerances for food, the agency proposed to prohibit use of PCBs in food, food-packaging and feed-manufacturing plants. A year later, the regulation was not in effect.

The late Congressman William F. Ryan of New York, in the wake of the North Carolina chicken contamination, introduced a bill to ban all sale of PCBs in the United States. Ryan, who had been concerned about PCB contamination for several years, said, however, that responsible agencies could ban PCBs on their own. He said that prior to the North Carolina PCB accident he had repeatedly urged the FDA, the USDA, the EPA and the Secretary of Interior to ban PCBs. The incident at Holly Farms, said Ryan, "tragically illustrates the failure of the federal government to take preventable action against the unconscionable use of this deadly chemical."

Most disturbing, however, are the government's proclamations that there is no problem. FDA officials at a New York conference in 1972 assured everyone that the PCB problem was "under control." When sixty thousand PCB-tainted eggs got loose in Washington, the FDA's public-information officers went to great lengths to tell the public no one could get sick from eating the eggs (which was true in *acute* terms, but not over the long term). In its proposed PCB order, the FDA stressed that "PCBs are not considered to be an immediate hazard to public health," although "PCB levels in food and animal feed must be reduced to minimize the longer-term exposure."

There is a problem, and that fact should not be hidden from the public. The public should not be treated as children who cannot be alarmed. If truly informed, the public can be counted on to take far stricter views on correcting environmental hazards than the conglomerate of federal agencies.

Remember, too, that when government officials say there is no human health danger from PCB-contaminated food, they do not necessarily mean PCBs are not present. The double-talk is illustrated by statements from Agriculture's Dr. Yeutter after the incident at Holly Farms. At first Dr. Yeutter insisted there "was no evidence that any contaminated chickens have reached any consumer." When called on that, he explained the contradiction with: "There was no statement that chickens fed on the contaminated fish meal had not reached the market. The statement was that there was no evidence—and there still is none—that any of the marketed chickens contained more PCB than the 5 parts per million *considered acceptable by the government.*" (Italics added.)

Thus, contamination is not considered as such when it is sanctioned by the government. As James S. Turner succinctly pointed out in *The Chemical Feast*, the public's standard of safety and the government's may be two different things. Wrote Turner: "When the FDA says a substance is safe, it merely means that for its own reasons—usually not explained —it has chosen to take no action against the substance."

This brings us to another kind of unintentional contamination—not by synthetic chemicals but by naturally occurring substances. No book on food safety would be sufficiently enlightening without mention of those most repulsive incidental additives: cockroach legs, rodent droppings, mold, insect eggs, worms, larvae, etc. For years the FDA kept secret from the public its standards of what constitutes filth—though they were known to some manufacturers. In March 1972, under pressure from consumer advocates and members of Congress, the agency finally released its "filth tolerances," which it calls "unavoidable defects" that "present no health hazard."

To a public that thought it was getting pure food, the report was a shocker. The FDA said it generally did not take legal action against a product unless it exceeded the following limits: chocolate, sixty microscopic insect fragments per 100 grams; cornmeal, an average of one whole insect per 50 grams; raisins, ten insects per eight ounces; wheat, one rodent pellet per pint; apple butter, an average of four rodent hairs per 100 grams; peanut butter, an average of fifty insect fragments per 100 grams; canned or frozen asparagus, 15 percent of the spears infested with six attached asparagus beetle eggs or egg sacs; frozen brussels sprouts, forty aphids or thrips per 100 grams; canned or frozen spinach, two spinach worms (caterpillars) five millimeters in length in twelve No. 2 cans; canned tomatoes, ten Drosophila fly eggs per 500 grams. (For a complete list of the filth tolerances, see the Appendix.)

According to the FDA, these tolerances exist "because it is not now possible, and never has been possible, to grow in the field, harvest and process some crops that are totally free of natural or unavoidable defects." Though the FDA admits that such filth is "aesthetically unpleasant," it assures us it presents "no hazard to health." That is not what the agency was saying a few years ago in its book *General Principles of Food Sanitation* (1968 edition). The agency specifically warned against filth, noting that rodents and flies carry disease bacteria, constituting a "real health hazard," and that insect bodies may carry "untold numbers of organisms of disease." To quote the FDA:

Often the line of demarcation between a harmful and filthy food is exceedingly narrow. Many sources of filth in food products are potential sources of disease organisms. Flies and roaches may harbor pathogenic bacteria and transmit infections to foods. Rodents, flies and other insects closely associated with filth and unsanitary conditions are capable of mechanically transferring pathogenic and spoilage organisms from such filth directly to food products.

The filth tolerances were first set in 1911 and were revised in the 1930s and again in 1973 as a result of howls of public protest. Still, the tolerances are outrageously high and allow marginal manufacturers to continue unsanitary practices that other manufacturers easily eliminate. In a scathing article on the government's filth standards, *Consumer Reports* magazine, published by Consumers' Union, termed the FDA's current attitude about filth in food "grossly irresponsible." Observed CR: "We find the FDA's standards grossly out of line with current food processing capabilities." The magazine pointed out that, contrary to FDA assertions, many types of contamination are avoidable. "High levels of filth often result from using infested raw materials. Rat excreta or roach fragments in food are signs of improper storage or manufacturing practices." CR noted that 86 percent of the peanut butter samples it tested in 1972 contained *neither* insect fragments nor rodent hairs. These are patently preventable infestations. The FDA's allowance for rodent excrement in wheat (up to one pellet per pint), said CR, "does not even reflect average conditions found more than two decades ago, and conditions have improved since then."

The same situation exists with corn, said the magazine. Corn used to be stored in corn cribs vulnerable to rats and mice. "Such storage conditions are comparatively rare now, but the FDA filth standards that were adopted to correct the situation that existed in 1940 have weathered the generation gap." Many manufacturers now have much more rigid guidelines against filth than the FDA. "By setting such high tolerances for filth, the FDA is, in effect, encouraging adulteration of foods rather than assuring their purity," concluded CR.[7]

By setting an unnecessarily high guideline for another natural contaminant, aflatoxins, the FDA is also exposing Americans unduly to disease. Aflatoxins are a group of potent poisons produced by *Aspergillus flavus* molds, which can form under humid conditions on such foods as wheat grain, corn, and nuts of all kinds. The poisons were discovered in 1960 when the death of twelve thousand young turkeys in England was traced to feed containing peanut meal that was contaminated with aflatoxins. Since then aflatoxins have proved to cause a fatal disease in dogs, called hepatitis-X, and cancers in rats and other species in which they have been tested. One aflatoxin, designated Aflatoxin B-1, is regarded as the most potent carcinogen ever found; a mere fifteen parts per billion fed to rats cause liver cancer in 100 percent of them.[8] Epidemiologists have also linked aflatoxins to human illnesses in Thailand, the Philippines, South Africa, Uganda and India. Studies in Thailand show that people eating food contaminated with aflatoxin have a high incidence of liver cancer. In India, after twenty malnourished children were accidentally given a protein supplement contaminated with three hundred parts per billion of aflatoxin, three died of cirrhosis of the liver within eighteen months.

Despite aflatoxins' carcinogenic potency, the FDA takes no action against food contaminated with less than twenty parts per billion. Not long ago, an FDA official justified this "guideline," saying, "Though the Delaney clause allows no tolerance for a known carcinogen, we stick to the 20 ppb. guideline. We can't accurately measure much less than that."

As long ago as October 1971, chemists at a meeting of the Association of Official Analytical Chemists announced they had improved methodology so that aflatoxins could be reliably detected at five parts per billion. In fact, there was wide speculation that the FDA would immediately lower the tolerance 300 percent to five parts per billion in line with improved technology, as Canada has done. But FDA officials decided to stick with the twenty parts per billion. Furthermore, even five parts per billion is not a floor. Consumers' Union analyzed peanut butter samples for aflatoxin at two parts per billion, and the FDA has a method sensitive to one part per billion for finding aflatoxin residues in the flesh of animals fed aflatoxin-contaminated feed.

The twenty-parts-per-billion tolerance is simply a figure the

FDA uses for economic purposes. And it subverts the aim of keeping carcinogens out of foods. Such was the case with some cornmeal mixes analyzed at the FDA; they were contaminated at nineteen parts per billion of aflatoxin, a dangerous dose, but since this squeaked under the legal limit, they were not touched by the FDA and were shipped out and consumed by Americans—all under a "technicality" of the law. (Since aflatoxin is invisible and tasteless, there is no way consumers can detect it. Even government chemists can spot it only through chemical analysis, for it may be present even though there are no obvious signs of mold.)

Yet the FDA's Dr. Wodicka has called aflatoxins one of the agency's most worrisome problems and a national hazard. He has acknowledged that "although levels sufficient to cause acute poisoning are extremely unlikely, the regular consumption of these materials over an extended period can lead to liver cancer, and there is considerable indirect evidence that it does in some countries."

Food becomes contaminated with aflatoxin mainly because of poor storage practices which allow grain to become damp, thus breeding molds. Since aflatoxin is potently toxic to chick embryos, causing early deaths and deformities, one method of confirming its presence is through the chick-embryo test. I have analyzed many samples sent to me from FDA field men throughout the country. In 1971 aflatoxin contamination was found in excess of twenty parts per billion in 140 food samples—notably in corn, cornmeal, hominy feed, pistachio nuts, cottonseed, copra (coconut) meal, cornmeal mixes. The FDA has seized or recalled corn, cottonseed, almonds, cornmeal mixes, but in many instances the contaminated grains or nuts have been sold because the dealer would not voluntarily withhold the product while the FDA was doing the analysis, and the FDA cannot make a seizure without proof of contamination. Confronted with the guideline, some offenders have also mixed their highly contaminated grain with less contaminated grain to reduce the total contamination level below twenty parts per billion.

There is no reason—except economic—that the aflatoxin tolerance could not immediately be reduced at least to five parts per billion for so potent a carcinogen. Nor should grain contaminated above that level be allowed to be fed, as it is now, to milk- or meat-producing animals, when the aflatoxin can show up in food intended for humans. We don't want aflatoxin-contaminated livers any more than we want DES-contaminated livers.

The necessity for low tolerances also applies to all kinds of pesticides and herbicides, such as 2, 4-D and 2,4,5-T, which contain the dread dioxins, known to cause terrible birth defects in animals at exceedingly low levels. (Only 125 to 500 parts per trillion fed daily have caused deformities in rats.) These herbicides were used as defoliants in Vietnam and are still being widely used in the United States; consequently dioxins do show up in food as a result of irresponsible spraying. So life-threatening are these defoliants that they should be immediately and totally banned from use, as some scientists have recommended.

Consumers should also be on the lookout for increased contamination from another incidental additive: ionizing irradiation of food. Even now the FDA permits irradiation of wheat for disinfestation, to kill or sterilize bugs, and irradiation of potatoes and onions to prohibit sprouting, though it is not yet widely used for these purposes. The use of irradiation on strawberries, fish and other perishables has also been proposed to extend their shelf life; irradiation changes food chemically, retarding deterioration. Bacon subjected to high doses of irradiation can remain unrefrigerated for two years, but such usage is not approved. That such usage will be contemplated in the future seems inevitable; for example, astronauts took irradiated bread on their moon trips. Moreover, irradiation in food processing—for example, to achieve the form-fitting plastic packaging of turkeys—is now allowed. As a result, consumers are undoubtedly being subjected to the byproducts of irradiation in food.

Irradiation at high levels has been shown not only to severely destroy vitamins and minerals in food, but also to cause reproductive problems, a shortening of the life span and other complications in laboratory animals. In some instances —for example, in irradiated jams and fruit compote—cancer is a suspected result.

Incidental additives, then, can be fully as dangerous as intentional ones, and we must do everything possible to reduce them to their lowest levels—instead of setting high legal tolerances that condone and encourage adulterated foods.

First Cyclamates, Now Saccharin

Cyclamates, the synthetic sweeteners, have passed from the American supermarket shelf, though occasionally you hear reports that the industry still hasn't given up. McCormick's Richard L. Hall, former president of the Flavor and Extract Manufacturers' Association, said in May 1972 that "the cyclamate story is far from over." Believe it or not, cyclamate and its metabolic byproduct cyclohexylamine (CHA) are still being tested, and in June 1972 Dr. Oser's Food and Drug Research Laboratories announced that CHA had come out clean in new tests. Some observers speculate that there is an underground movement afoot to "reevaluate" cyclamates as evidence mounts against saccharin, the artificial sweetener that industry fell back on when cyclamates were banned.

When asked about the safety of saccharin, an FDA official, in October 1972, called it "a story in the making." He failed to add that it's the same old story, with nearly identical details, as the one the public heard about cyclamate. At present it's too soon to tell whether the saccharin story will have a surprise ending—or, for that matter, whether cyclamate will have a comeback. But it is instructive to look at the FDA's handling of cyclamate and note how little we've learned when the scenario can repeat itself.

The cyclamate incident was one of the biggest explosives to hit the FDA. Officially, you might say it began to sizzle on October 1, 1969, when, after clearing the matter with superiors, I was interviewed on NBC television. I stated that in my experiments injecting cyclamate and CHA into about thirteen thousand chick embryos, I had found that both caused grotesque birth abnormalities, such as twisted spines and legs and phocomelia—the dreaded stunted limbs and flippers common in thalidomide babies. Actually, these abnormalities were worse than any I had seen in similar experiments with thalidomide.

Considering the severity of the situation, my comments on television seem, in retrospect, cautious and fairly innocuous. When asked what my experiments meant to pregnant women, I replied: "It would seem that at least here an effect on reproduction has been demonstrated. It's only in the chicken embryo. However, I think the prudent thing is to avoid anything which is unnecessary and which is not given under medical supervision perhaps, until further information has been gained." For some incomprehensible reason my mild statement was taken by some in the agency and in HEW as highly inflammatory, and I was accused of a breach of etiquette in "going directly to the media without having consulted my superiors" (though I had consulted them). Surgeon General Jesse R. Steinfeld criticized my action, as did then HEW Secretary Robert Finch, who publicly termed it "unethical."

My revelation should hardly have been a bombshell. Evidence that cyclamate should be stricken from the GRAS list had been circulating around the FDA and the scientific community for fifteen years. And even industry spokesmen admitted it wasn't safe for pregnant women. A month prior to my TV appearance, Dr. Claire Dick, a cytogeneticist with Abbott Laboratories, the largest producer of cyclamates, had told a Pittsburgh TV audience, when asked by an interviewer whether she would eat cyclamates if she were pregnant, "No, I don't think I would."[1] I had written up my findings in numerous staff reports, specifically in my year-end report for 1968; they had been discussed at internal meetings, and early in 1969 one of my supervisors had taken cyclamate-deformed chicks to Commissioner Herbert Ley's office to impress upon him the hazards of this chemical, but somehow they were overlooked in the rush of other business.

In November 1968 Dr. Marvin Legator at the FDA reported publicly that CHA was a mutagen, producing broken chromosomes in both test-tube experiments and live animals. Moreover, he believed, the doses were not much more than those being consumed in a variety of foods and especially in carbonated diet beverages. In a memo to the commissioner two months later, he urged: "The use of cyclamates should be immediately curtailed, pending the outcome of additional studies." But this sentence was deleted by Legator's superiors without his knowledge, and the commissioner never saw it.

When they first came into use in the 1950s as artificial sweeteners, cyclamates were intended only for consumption on a restricted basis by those with medical reason to curtail their sugar intake. Then in the 1960s the situation got out of hand. Yearly cyclamate production increased fivefold—to 12 million pounds in 1966 and 17 million pounds in 1968. At the time of its ban it was in $1 billion worth of food. You found it in Tab and Diet Pepsi, in canned fruits, jellies, jams, ice cream, iced-tea mixes, Kool-Aid, in the coating of children's vitamins and even as a curing agent in bacon. Weight-conscious Americans loved it because they thought it kept them thin (though there is no evidence it did), and the food manufacturers loved it because it gave them a selling tool and saved them millions of dollars. For only sixty-four cents' worth of cyclamate they could replace six dollars' worth of sugar!

As early as 1954, a National Academy of Sciences Food Nutrition Board voiced doubts about cyclamate's safety. As its uncontrolled use soared (despite warnings on the labels) and as evidence piled up strengthening the doubts, private researchers began asking the FDA to take cyclamate off the GRAS list, and memos urging that action were flying around the FDA.

September 8, 1967 (a memo from an FDA toxicologist to a superior): "We cannot say today that the cyclamates are generally recognized as safe; however, removing them from the GRAS list and establishing tolerances in soft drinks, etc., will produce difficult problems."

October 30, 1967 (a memo reporting a meeting of FDA officials with researchers from the Wisconsin Alumni Research Foundation—WARF—who had been conducting studies on cyclamates): "They felt the cyclamates should clearly be removed from the GRAS list. Their first proposal was to limit the product to use in truly special dietary areas where nutritive sweeteners are contraindicated, but after further discussion of the possible harmful effects they believe have been shown they changed their position to the point where the cyclamates should be ruled out of our food supply completely."

December 4, 1968 (a memo from a toxicologist to his superior): "The cyclamates should be removed from the GRAS status. They should be considered to be food additives and tolerances could be established to limit the amount according to the usage of the product."

December 13, 1968 (from the Associate Commissioner for Science to the Deputy Commissioner): "Cyclamates will be removed from the GRAS list . . ."[2]

A major spur to alarm was the discovery by the Japanese in 1966 that cyclamates could metabolize in the body to CHA, long known to the FDA as a dangerous chemical. It was later found that about one fifth of the population converted cyclamates to CHA, and in some instances CHA had already formed in products offered for sale—so no one was immune from harm. There was a string of other evidence against cyclamate: It affected the intestinal tract, causing diarrhea and softening of the stool in humans. It caused liver damage in guinea pigs. It interfered with the body's absorption of certain drugs. It was thought to interfere with the body's absorption of vitamin K, which might lead to bleeding problems. CHA appeared to have a greater toxicity to fetuses and newborn infants, and there was clear evidence that cyclamates were able to cross the placenta.

There was ample reason to take cyclamates off the GRAS list, even to ban them. The FDA did neither. Instead, it referred the matter to a National Academy of Sciences/National Research Council ad-hoc committee, chaired by Dr. Coon—which in November 1968 came up with an interim report concluding that though there was some damaging evidence and cyclamates shouldn't be "totally unrestricted," it was all right for adults to go on eating them as long as they didn't exceed five grams of the stuff per day. What studies they had to back up this "safe dose" were not cited. Many at the FDA thought the report shoddy (even the general counsel, William Goodrich, later called it "disappointing"), but the FDA went along with it and, in April, called for firmer labeling, advising a maximum daily safe dose of 3.5 grams for adults and 1.5 grams for children on the basis that higher levels could cause diarrhea.

But asking people to regulate their intake is an absurd control, for how many people are going to go around with pencil and paper adding up how much cyclamate they're eating every day? Besides, it was hopeless, since the cyclamate content of some of the products was so excessive (Kool-Aid was 30 percent cyclamate) that you could quickly use up your daily quota. One package of Kool-Aid, which some youngsters ate in powder form, contained more than the daily recommended safe dose for kids, and some other beverage powders contained two and a half times the 1.5-gram dosage.

By June 1969, Dr. George T. Bryan and associates at the University of Wisconsin Medical School came up with significant studies showing that pellets of cyclamate implanted in mice caused bladder cancer. Though the findings were shunted off by some as meaningless (because the mice hadn't been *fed* the cyclamate), they were highly predictive. Industry at least felt it had enough warning signs to begin looking seriously for replacements. Pepsi-Cola, in fact, had a new diet drink without cyclamates ready in late 1968, and Coca Cola said it had taken out "insurance" that year by working on substitute diet drinks against the day when cyclamates would go down the drain.

The climax came on October 11—ten days after my TV appearance—when Abbott Laboratories informed the FDA that the Food and Drug Research Labs had found cancers in the bladders of rats fed Sucaryl, a mixture of nine parts cyclamate and one part saccharin. When these data were confirmed by data from National Cancer Institute scientists, the FDA hastily reconvened the NAS committee on cyclamates and, on October 18, with the concurrence of both groups, reluctantly ordered cyclamates off the GRAS list.

Secretary Finch stated that cyclamated "general-purpose foods and beverages" such as the diet drinks would be off the market by January 1 (a deadline that was later moved to April 1), and that other foods with cyclamate, such as diet canned fruits, would be prohibited after February 1 (a deadline that became September 1, giving fruit canners nearly a year to sell off their cyclamate-laden stocks). FDA general counsel William Goodrich later explained to a Congressional committee, "There was a very substantial volume of these canned fruits and vegetables just hitting the market when the October announcement was made."[3] The canning season, he went on to say, had just ended in California—which, coincidentally, was Finch's home state, as well as President Nixon's.

Now, one might have thought that the government was banning cyclamates as required under the Delaney clause. But there was an almost unbelievable catch. As Finch put it that day: "I should emphasize also that my order does not require the total disappearance from the marketplace of soft drinks, foods and nonprescription drugs containing cyclamates. These products will continue to be available to persons whose health depends upon them, such as those under medical care for such conditions as diabetes or obesity. I expect that in the future these products will be labeled as drugs, to be consumed on the advice of a physician."

In short, the government didn't plan to ban cyclamated foods at all; it planned to declare them "over-the-counter drugs" and allow them to be sold in grocery stores as before, but with a stricter warning label. It was a high-level decision that emanated from the Secretary's office. Then the FDA was faced with the problem of how best to perform this magical trick of transforming food into drugs with some semblance of legality. The task of providing justification for the move fell to Dr. Roger Egeberg, HEW Assistant Secretary for Health and Scientific Affairs, himself a portly man, who said he took cyclamates to keep his weight down and advised others to do the same. A couple of weeks after the ban he told a meeting at the Harvard Medical School, "I'd even advise my daughter to stock up a bit on diet drinks which will soon disappear from the grocery shelves by federal order, as long as she drank only a bottle or two a day."[4]

Dr. Egeberg convened a hand-picked "medical advisory group" of outside physicians, who not unsurprisingly came to the conclusion that cyclamates, though "not absolutely necessary in any disease, can be useful in the medical management of individuals with diabetes and patients in whom weight reduction and control is essential to health." They recommended that it be available under "medical supervision" on a "nonprescription drug basis."

Lawyer Goodrich, bending to the wishes of the high command, then had to find a way to justify what was a *fait accompli*. He asked food manufacturers to submit "abbreviated new drug applications" for their cyclamated foods—such as jellies, jams, fruits and vegetables. Big and little executives alike whose previous concern had been solely with food now found themselves in the drug business. It was the FDA's greatest moment of low comedy, for everyone knew that this was a move to circumvent the Delaney clause. It was not only foolish but dangerous. The FDA had never before, as Goodrich admitted publicly, approved a known cancer-causing substance for over-the-counter use.

There were a few other drawbacks. Nobody could dig up evidence that cyclamate was "essential" for the medical management of diabetics and the obese, or even that is was "beneficial," which a new drug must be. Mary McEniry, a lawyer in the FDA's Bureau of Drugs, thought the whole idea absurd. She wrote in a memo dated January 30, 1970: "To approve cyclamate-containing foods as safe and effective drugs within the concept of our enforcement of the new drug provisions of the act is untenable. We are aware of no evidence that cyclamate-containing foods are safe or effective for use in the treatment of obesity or diabetes. Under the principles we strongly adhere to in permitting drugs to be marketed, these products should not be allowed on the market. To approve an NDA [new drug application] for these products is not supportable medically or legally."[5] She suggested that if the FDA was going ahead with this notion, it should start exploring other ways of doing it.

The FDA's John J. Schrogie noted: "None of the few controlled studies reported to date have established a useful role for nonnutritive sweeteners as weight reducing except under the most carefully controlled circumstances." Dr. John Jennings, the FDA's Associate Commissioner for Medical Affairs, confessed to the Fountain committee that they didn't have the "substantial evidence for safety and efficacy" that they usually required for routine approval of new drugs. Indeed, a well-controlled study done at Harvard with diabetics and obese had found no weight-loss difference between those who used artificially sweetened foods and those who didn't. A report in *Nature* even showed that cyclamates stimulated the appetite of rats, causing them to gain more weight and use food more efficiently.

Eventually, the idea of classifying cyclamates as drugs vanished under pressure or was laughed to death. By the time the September 1970 deadline for declaring such products drugs rolled around, the FDA declined to go through with it. To save face, HEW hastily reconvened Egeberg's "medical advisory group," which, in the light of "additional" evidence, withdrew its recommendation. (Actually, the "additional" evidence cited—mainly a new study at the FDA showing that rats developed cancer when they were fed only one sixth as much cyclamate as in the previous Food and Drug Research Lab tests—had been available to the group the first time around.) So, as of August 27, 1970, cyclamates were totally banned.

They were not in the public eye again until 1972, when former cyclamate producers persuaded members of Congress to introduce bills to compensate them for their losses due to the ban, to the tune of from $100 million to $500 million of taxpayers' money. The effort was spearheaded by the major supplier of cyclamate, Abbott (supported by the National Canners' Association and the Departments of Agriculture and Commerce), which claimed that the ban had taken it by surprise. Commissioner Edwards, who opposed the indemnity (this time in unison with consumer advocates), countered that the companies had "sufficient warning that the storm clouds were picking up long before it was banned." The bill, obligating consumers to bear the economic as well as the health burden from the unwise use of cyclamates, did pass the House, but died in 1972 in the Senate Judiciary Committee. There was talk that it would be revived in a future session of Congress.

The cyclamate incident was the one that triggered renewed interest in food chemicals and led to a review of the GRAS list. It also set the stage for the attitudes that would permeate the FDA's handling of future food-additive problems. The cyclamate scenario, directed by Robert Finch, was one of misinformation, manhandling of legal and scientific matters, and resistance to enforcing the law. And it left the public confused over the basic issues in food-chemical safety. Finch set the tone in his October press conference announcing the elimination of cyclamate from the GRAS list. He virtually apologized for his action: "I have acted under the provisions of the law because . . . I am required to do so." He was described by science writer William Hines as "looking for all the world like Pontius Pilate calling for a washbasin and towel." Subordinates followed his lead, and it was easy to see that high-ranking officials in HEW mocked the job they were required to do. Secretary Finch even delivered a poem about his predicament to the National Press Club:

> Damage in brains, mutations and cancer—
> Who in the world can give us the answer?
> Chromosome breaks, chick malformation,
> Pot and pollution and desegregation,
> So revising Delaney should just be a cinch—
> But why in the hell must it always be Finch?[6]

Officials misled the public into believing the ban foolish. They labeled the doses of cyclamate required to cause animal damage "strong," "high," "massive," when in fact the harmful doses were relatively low, providing a slim margin of safety; in an FDA study, cyclamates caused bladder cancer at only eight times the human consumption level. The officials muddled scientific facts. Surgeon General Steinfeld used the "no human evidence" argument as reassuring to the public, and pointed to Connecticut figures on bladder-cancer decrease as proof. Besides the fallacy of the interpretation, his facts were wrong; bladder cancer in Connecticut had doubled in the last two decades. The Administration used the cyclamate incident to launch an assault on the Delaney clause which is still going strong. All in all, it put a face of frivolity and unconcern on a problem of grave public consequence.

Now the question of the safety of saccharin has come to public attention, and history is repeating itself. Saccharin, like cyclamate, is a nonnutritive sweetener; it was synthesized in 1879 from coal tar. It fell out of favor with the advent of cyclamates (because saccharin, unlike cyclamates, gave a bitter aftertaste) but returned as a replacement when cyclamates were banned. It has now replaced cyclamates in soft drinks, canned fruits and vegetables, ice cream, etc., as an artificial sweetener.

Saccharin too got on the GRAS list in 1958; yet it has a far worse history of hazard than cyclamate. For more than twenty years scientists at the FDA have urged further study of saccharin. In 1953, FDA pathologist A. A. Nelson, who did the classic study of artificial sweeteners, warned that he suspected saccharin might cause cancer and called for further experiments to confirm or disprove his fears. Nothing was done. But when his slides and research findings were reviewed in December 1970, kidney lesions (changes in normal cell structure) were found and were definitely linked to the saccharin intake. Studies I did on the chick embryo in 1970 showed that saccharin causes minor abnormalities, such as beak and eye deformities, and in higher doses produces baby chicks that are bleached snow white, showing interference with metabolism. More significant, in some experiments as many as 30 percent of the saccharin-injected eggs produced chicks that died within forty-eight hours after hatching.

That same year, Dr. Bryan, after tests, warned that "people ought to be alerted that . . . all is not well" with saccharin. Dr. Bryan found that when he implanted pellets containing saccharin into the urinary bladders of mice, they developed cancer at high rates. In one test on a group of sixty-six mice, cancer occurred in 47 percent. In another group of sixty-four, it was present in 52 percent. Some scientists disparage the pellet-implant technique, since the route of exposure is not the same as in humans. However, in the case of cyclamate the pellets were an accurate predicter, confirmed by later feeding studies. Cancer researcher Dr. Saffiotti terms the pellet method "a sensitive experiment, an extremely useful warning signal and a good indicator." He said in 1970 that on the basis of Dr. Bryan's work there was "a good chance" saccharin would turn out carcinogenic, though feeding studies were needed for conclusive proof.

Actually, Dr. Bryan's findings were not new. The FDA had in its files reports of similar studies done in England in 1957. But pushed into action by Dr. Bryan's public disclosures, the FDA called on the National Academy of Sciences to review the matter. An NAS/NRC panel reviewed the information on saccharin and came up with a predictable conclusion, almost identical to the one they had come up with on cyclamate. First, the report restated a fact which everyone knew and which had no relevance at all—that saccharin is of very low acute toxicity. Studies in diabetics and children, the report said, "tend to confirm the evidence from animal studies that saccharin has a wide range of safety for acute and short-term exposure." There was no mention that millions of youngsters and adults were consuming saccharin, especially in soft drinks, over a *long* period of time and would continue to do so without government intervention.

The evaluating panel also pointed to the long, apparently safe usage of saccharin, noting: "Although the 80-year history of saccharin use by man without evidence of adverse effects . . . cannot be accepted as final proof of its safety for chronic consumption, it should be given due consideration in the over-all toxicological evaluation." This is the kind of scientific gibberish typical of these committees. Why in heaven's name should we give weight to the fact that in eighty years of eating saccharin we haven't dropped dead from acute saccharin poisoning? It is not only *not* "final" proof of lack of chronic toxicity; it is no proof at all. The Academy committee concluded that the use of saccharin in the United States does not "pose a hazard."

By early 1972 new evidence was piling up against saccharin from feeding studies. Researchers at the Wisconsin Alumni Research Foundation notified the FDA that they were finding tumors, some malignant, in rats fed high dosages of saccharin, about 5 percent of their diet. In February 1972, the FDA, in one of the tardiest gestures ever made, removed saccharin from the GRAS list, but did not prohibit it; like other suspect additives, saccharin was thrown into the never-never land of the "interim food additive regulations," and industry was given until June 30, 1973, to "conduct additional studies."* It was easy to see that saccharin was following the same path as cyclamates: first the data from the 1951 studies, then from the pellet implants in Wisconsin causing bladder cancer, then the "reviews" by the National Academy, finding no need for alarm, then the evidence from feeding studies.

There were rumors within the FDA for some time that experiments with saccharin in FDA labs were showing bladder tumors in rats. In February 1973 the FDA officially announced it. Said a press release: "FDA technicians report they have found bladder tumors in rats fed saccharin. In the FDA study, rats were fed saccharin for up to two years in doses from 0.015 to 7.5 of their total daily diet." The release then went on to disparage the data and destroy public concern by noting that "7.5 saccharin in a rat's diet is roughly equivalent in humans to 1,300 bottles of a typical diet soft drink per day." It was a typical misleading statement.

FDA officials further stated they did not know whether the tumors were cancerous (it's amazing how, in humans, pathologists can determine this in minutes while a patient is still on the operating table, but the FDA apparently needs days or weeks) or whether they were caused by impurities in the saccharin (which is irrelevant, since the tests used the table-grade saccharin eaten by humans). In any event, the FDA stressed it would take no immediate action, but would await completion of its studies, as well as others in progress, and would then refer the matter once again to the National Academy of Sciences' ad-hoc committee on artificial sweeteners, chaired by Dr. Coon. Said the FDA's Dr. Wodicka, "We're not going to take any action until we get a recommendation from the National Academy of Sciences." This

* On May 24, 1973, the FDA announced that it was extending the interim regulation again "until National Academy of Sciences advice can be received." "It is clear," said the FDA in a press release, "that the Academy will be unable to receive and evaluate data coming to it from the various studies in time to meet the original target date of June 30, 1973."

time there were no big press conferences, and no quick action to ban the last of the artificial sweeteners.

The FDA had learned the advantages of stalling, of drawing out the proceedings, so as not to cause a crisis in the food industry. Manufacturers were getting plenty of lead time to develop substitutes and reformulate products. At the time the cancers from saccharin were showing up, the FDA had received a food additive petition for a new artificial sweetener much sweeter than saccharin. However, this does not excuse the FDA's irresponsibility in allowing Americans to continue to consume a chemical which everyone at the FDA knows in his heart is unsafe and will have to be banned—if not today, then in the near future. As Congressman Delaney said about saccharin in early 1973, "To needlessly expose millions of people to a potentially harmful chemical during a number of years when its safety is being tested is totally unwarranted. Public health must not be sacrificed upon the altar of profit and economic expediency."

The harm that occurs during these grace periods is inestimable. Dr. Lederberg estimated that if cyclamate were as toxic to humans as to laboratory rats, it might have caused a million cases of human cancer during the decade it was used. Saccharin has been used much longer, yet the possibility of its harm does not seem to be a gut issue with those in charge. When asked in 1970 how he would feel if he knew that his failure to ban saccharin immediately was subjecting countless persons to cancer in the future, Dr. Charles Edwards looked pensive for a moment, then replied, "I just don't know."

SUMMING UP

One can only say that the FDA's performance, certainly of late, has been outrageous. The purpose of this book is to make that point clear, with the hope that consumers as well as legislators, often ill-informed about the state of the FDA, would get the message. This book, then—an unauthorized, unofficial report on the FDA—is a "countervailing" opinion, if you will.

Obviously, the final question is: What can be done to restore to consumers their right to safe food regardless of economic and political interests? Anyone who has seen the FDA organized and reorganized and scrutinized by Congress knows this is a tough one. Probably the best solution, as some members of Congress have suggested, is to abolish the FDA and start over with a completely new agency free of some of the political pressures. Sponsored by Senator Warren G. Magnuson, a bill calling for such an independent agency did pass the Senate in 1972, but it did not get through the House. Under Magnuson's plan, food chemicals would be handled by an independent Commissioner of Food Safety, subject to Senate confirmation, who would serve a fixed term; no superior, including the President, could remove the commissioner because of dislike of his actions. Furthermore, those in the new agency could not take a job with the food industry for at least three years after leaving their government post, under a conflict-of-interest provision.

Ralph Nader has another excellent idea to keep administrators and policy-makers from being corrupted. He would write into law a provision that officials who display gross dereliction of their duties could be sued by citizens and would be subject to fines, dismissal and *criminal* penalties. This provision obviously would be a constant reminder to food-protection officials of precisely who their constituency is.

Senator Gaylord Nelson has introduced a number of bills which are aimed at strengthening laws on food additives, and which consumers should soundly support. One would expand the Delaney clause to require a ban not only on cancer-causing additives but also on those that cause reproductive damage or mutations in animals or humans as demonstrated by valid studies or other sound scientific evidence. Such a move would relieve the FDA of its discretionary powers concerning birth defects and mutations, and would require an immediate ban on Red 2 food dye, for example, since it clearly caused fetal damage in rats. This is only a sensible extension of the law in light of present knowledge that there is no "safety threshold" for mutants or teratogens.

Senator Nelson has also proposed a law to reduce the number of unnecessary food additives. He would make manufacturers prove that their food chemicals not only are safe and effective for intended use but also "have demonstrable benefit or are unavoidable by good manufacturing practice." Says the senator, "The chemical should serve a socially and economically useful purpose for the general population. If it does not, there is no reason to risk potential hazards without matching benefits. There is no reason to introduce additional chemicals into the food supply when already approved, safe and effective alternatives are available . . . I believe that the number of chemicals used for food processing can be drastically reduced to a few effective substances."

And to keep testing honest, Nelson has proposed objective "third-party" testing. Presently, industry conducts its own animal tests or it contracts with so-called independent laboratories for tests on food chemicals. All too frequently, as has been noted, the test results conform to the wishes of the payor. It's hardly likely that industry will continue to plow money into a firm that consistently comes up with results showing that a certain food additive is not fit to eat. Thus, legitimate, hard-working, honest firms are caught in a bind and may subtly or perhaps unconsciously try to weight the studies to come up with a preconceived verdict of safe. To eliminate such prejudice, Nelson would funnel industry's money through the FDA; the agency would then decide which private laboratory should do a particular study financed by industry, and would approve the protocol. Industry, though paying for the experiments, would have no control over the choice of a lab, and the lab would be in no danger of not receiving future contracts should it turn up damaging evidence. Others would go further and would entrust the "third-party" testing choice to an independent board of distinguished scientists, beholden neither to FDA nor to industry. Under such a system, we should expect to get truly objective results and eliminate the endless testing which impedes the FDA's actions.

Certainly organized consumer pressure is a must—and consumers should morally and financially support groups such as Ralph Nader's Center for Study of Responsive Law and Health Research Group; Dr. Jacobson's Center for Science in the Public Interest; Jim Turner's Consumer Action for Improved Foods and Drugs; Consumers' Union; Ruth Desmond's Federation of Homemakers, which consistently watch the activities of the FDA and the USDA and bring lawsuits and other pressure against them in the public interest, to get dangerous chemicals removed from the market.

Consumers can organize locally on special issues, as a group has done in Lexington, Kentucky, to encourage local bakers to make nutritious additive-free bread. A Philadelphia organization, the Consumers' Education and Protective Association, picketed local supermarkets after filth in frozen breakfasts was found by Consumers' Union, and forced removal of the contaminated food from shelves.

Consumers should fight for complete labeling so that *all* additives will be noted. You should also refuse to buy foods known to contain suspect additives and should inform government agencies and the manufacturers of the reason for your boycott. Nothing encourages manufacturers to make changes or look for substitutes quicker than a drop in sales. If you hear that the FDA or the USDA is "reexamining" an additive, conducting new tests or referring the matter to the National Academy of Sciences, you can be sure there is much more evidence beneath the surface than has been revealed, and it would be wise to avoid the additive if possible. For by the time the FDA makes a public statement hinting at danger, it is a sure bet there is plenty of adverse evidence.

Consumers would also be well advised to buy fresh foods instead of the additive-loaded convenience foods. As long as many of these additives remain untested, people who eat them are simply engaging in a game of Russian Roulette. Furthermore, the highly processed foods are not usually as nutritious as fresh foods.

It was never the purpose of this book to lay out a legislative blueprint for reforming the FDA, but rather to make people, including members of Congress, aware of the scientific issues, the dangers and the lack of protection by agencies charged with saving us from unsafe food. The nature of the needed reforms is obvious, and if our Congressional representatives would vote into law the many measures already proposed to reform the FDA we would be a long way on the road to safer food. They will do it more speedily, of course, if they have an alert and aroused constituency. Consumers, both alone and organized, can and should exert political pressure—through letters, telegrams, lobbying—for food free from dangerous additives.

HOW TO ARGUE WITH YOUR GOVERNMENT: A PRIMER

Your government assumes you have little of the basic information necessary to dispute its pronouncements or decisions on food additives. Thus it is invariably left with the last word. We hope this book has given you essential information with which to talk back. But we also thought it might be helpful to put together in dialogue form the excuses given most often by government for its failure to protect you against harmful additives, and the appropriate arguments to use in response. Here's a dialogue in which the citizen for once has the last word:

CITIZEN (speaking to the FDA): Why do you allow such and such a chemical to remain in food when there are animal tests showing that it causes harm?

GOVERNMENT: We don't feel there is sufficient proof to find the chemical dangerous.

CITIZEN: Under the Food Additive Amendments of 1958, the government does not have to assume the burden of proof of finding an additive *unsafe*; manufacturers using the chemical must give sufficient proof that it is *safe*. By leaving the chemical on the market you are failing to uphold the law.

GOVERNMENT: It would be rash and unnecessary to remove it, because there is no "imminent hazard" to health. If we felt there was, we would end use of the chemical.

CITIZEN: What is "imminent hazard" to health? A chemical believed to cause cancer is always an "imminent hazard" to some persons. So are certain other food additives. If a chemical is a hazard six months from now, it is a hazard *today*. By not insisting on an immediate withdrawal, you are granting industry a grace period in which to deplete its stocks and find substitutes. It is true that the courts have upheld your interpretation of "imminent hazard" as something *acutely* toxic, but this concept is outmoded and ineffective to cope with the threat of today's chemicals. If the language "imminent hazard" does not provide sufficient protection for the public, the FDA should seek a change in the law. Protecting the consumer is the FDA's function; protecting business is a function of the Department of Commerce.

GOVERNMENT: But we must consider a "benefit–risk" here. The harm to consumers of removing the additive might be greater than the benefits of leaving it on the market.

CITIZEN: Once again, the FDA has no authority to trade off health benefits and economic benefits. The intent of the law is unequivocal: the FDA is mandated by Congress, representing all consumers, to make decisions solely on the question of whether an additive is *safe*—not whether it has any economic benefits. Nor has the FDA any qualifications for making such benefit–risk determinations. Most of the additives used in food are unnecessary for health reasons, are superfluous and cosmetic. How can the FDA possibly weigh a cosmetic benefit against a health risk? The FDA is in no position to decide what risks consumers must take in return for ill-defined benefits from additives, when consumers through their elected representatives in Congress have decreed they do not want to accept any unnecessary health risks.

GOVERNMENT: Additionally, animal tests showing the chemical to be hazardous are inconclusive; other tests show it to be perfectly safe.

CITIZEN: We are not playing a "world series" with scientific information. Animal testing is imperfect, which is the precise reason we should take as an indication of human danger any valid tests that show harm. The tests are more likely to miss hazard than to spot it, as numerous scientists, including Dr. Joshua Lederberg of Stanford and Dr. Umberto

Saffiotti of the National Cancer Institute, point out. In Dr. Lederberg's words, "There is no basis for the illusion that a compound that can cause harm in other cells will not affect human cells."

GOVERNMENT: It is well known that almost any chemical if given in high enough doses to animals will cause damage such as cancer and mutations.

CITIZEN: That has been proved untrue; in fact, studies show that few chemicals when given orally do provoke either cancer or mutations. In a study done by Bionetics Laboratories for the National Cancer Institute, only 10 percent of 120 chemicals tested produced cancer.

GOVERNMENT: Even if a chemical does cause cancer in animals, there is no reason it cannot be safely used in food at exceedingly low levels which will do no harm to humans.

CITIZEN: Such a threshold of safety has never been found for carcinogens for their effects on the body are cumulative and synergistic. Leading cancer specialists throughout the world, including the National Cancer Institute and the World Health Organization, agree on that point. Those who believe a safe threshold can be set are decidedly in the minority and often have an economic connection with the food or chemical industry. The Delaney clause forbids adding known carcinogens to food and is based on scientific principles fully supported by cancer authorities.

GOVERNMENT: The Delaney clause should be abolished because it is too restrictive and prevents scientific judgments.

CITIZEN: The Delaney clause is our greatest protection, because it allows no economic or risk-benefit judgments to be made by the FDA on the grave question of carcinogens in food. On the other hand, it leaves room for competent scientists to determine whether the chemical in question actually caused cancer in test animals or humans; this is a scientific judgment.

GOVERNMENT: Even if animal tests reveal harm, there is absolutely no evidence that this chemical ever caused illness in any human being.

CITIZEN: This is undoubtedly your most fatuous and misleading argument, which at first sounds reassuring, but on dissection is fraudulent. First, did anyone try to search out human harm from chemical X? And second, how would they go about it? Because all of us eat thousands of chemicals daily under the same circumstances, it is foolhardy to think we can pin harm for any particular illness on a single chemical. It is equally foolish to wait for such illness to show up—since the illness is often irreversible and of epidemic proportions—instead of heeding danger warning signs from animal studies.

GOVERNMENT: But additives are simply chemicals; they are not all bad for you.

CITIZEN: Admittedly, some chemicals may be safe for humans, but you can't lump all chemicals together and call them safe. Take arsenic, for example. This is just another argument to cloud the primary issue that some additives are dangerous and we should detect them.

GOVERNMENT: Even natural foods, such as potatoes and spinach, may contain harmful toxins. Should we ban them too?

CITIZEN: We have never assumed that all natural foods are safe over the long term, because they have rarely been tested for chronic toxicity or reproductive harm. However, this is a specious argument to trip up the unwary. That certain natural foods may contain unknown toxins is no excuse for further loading up our environment with more untested, potentially dangerous chemicals which are not natural. We cannot refuse to correct a known or suspected danger just because other dangers may remain. It's like refusing to arrest one criminal because we can't find them all.

GOVERNMENT: Your attitude would virtually destroy our food supply.

CITIZEN: This is untrue. Many additives are purely cosmetic, unnecessary and redundantly used even when safe substitutes are available. We are not asking that all additives be abandoned, but that they be *tested for safety before use* as the law requires. All we're asking is that you do your rightful job of protecting us from harmful chemicals in food.

APPENDIX

Filth Guidelines in Food

As of January 1973

Following are the levels of natural contaminants (insects, etc.) that if found in food can cause the FDA to take legal action to remove the food from the market.

PRODUCT	DEFECT ACTION LEVEL
Chocolate and Chocolate Products	
Chocolate and Chocolate Liquor	Average of 60 microscopic insect fragments per 100 gm. when six 100-gm. subsamples are examined; or if any one subsample contains 100 insect fragments. Average of 1.5 rodent hairs per 100 gm. when six 100-gram subsamples are examined; or if any one subsample contains 4 rodent hairs.
Chocolate Liquor	If shell is 2% by weight calculated on the basis of alkali-free nibs.
Cocoa Powder, Press Cake	Average of 75 microscopic insect fragments per 50 gm. when six 50-gm. subsamples are examined; or if any one subsample contains 125 insect fragments.

PRODUCT	DEFECT ACTION LEVEL
	Average of 2 rodent hairs per 50 gm. when six 5-gm. subsamples are examined; or if any one subsample contains 5 rodent hairs.
	If shell is 2% by weight calculated on the basis of alkali-free nibs.
Cocoa Beans	If 4% by count show mold or 4% by count are insect-infested or -damaged; or if total of 6% by count show mold and are insect-infested.
	10 mg. of excreta per pound.
Coffee Beans	If 10% by count are insect infested or insect-damaged or show mold.
Eggs and Frozen Egg Products	
Dried Whole Eggs Dried Egg Yolks	Decomposed as determined by direct microscopic count of 100,000,000 bacteria per gram.
Frozen Eggs and Other	If two cans contain decomposed eggs; if subsamples examined from cans classed as decomposed have counts of 5,000,000 bacteria per gram.
Fish and Shellfish	
Blue-Fin and Other Fresh-Water Herring	Fish averaging 1 lb. or less: 60 cysts per 100 fish, provided that 20% of the fish examined are affected. Fish averaging over 1 lb.: 60 cysts per 100 lbs. of fish, provided that 20% of the fish examined are affected.
Rose Fish (Red Fish and Ocean Perch) Fresh and Frozen Fish	If 3% by count of the fillets examined contain one copepod. If (1) 5% by count of fish or fillets in sample (but not less than

PRODUCT	DEFECT ACTION LEVEL
	5) show Class 3 decomposition* over 25% of their areas; or
	(2) 20% of the fish or fillets in the sample (but not less than 5) show Class 2 decomposition over 25% of their areas; or
	(3) The percentage of fish or fillets showing Class 2 decomposition as above, plus 4 times the percentage of those showing Class 3 decomposition as above, equals at least 20% and there are 5 decomposed fish or fillets in the sample.
Fresh and Frozen Tullibees, Ciscoes, Inconnus, Chubs and Whitefish	50 cysts per 100 lbs. (whole fish or fillets), provided that 20% of fish examined are infested.
Flours and Cornmeals Cornmeal	Average of one whole insect (or equivalent) per 50 gm.; or average of 25 insect fragments per 25 gm.; or average of one rodent hair per 26 gm.; or average of one excreta fragment per 50 gm.
Fruit Apricots (canned)	Average of 2% by count insect-infested or -damaged.
Caneberries (blackberries, raspberries, etc.—canned or frozen)	*Frozen black raspberries:* microscopic mold-count average of 60%. Average of 4 larvae per 500 gm.; or average of 10 insects (larvae & other insects) per 500 gm. (excluding thrips, aphids and mites).
Cherries (brined, fresh, canned or frozen)	*Brined and Maraschino:* Average of 5% rejects due to larvae.

* Definition of classes of decomposition:
Class 1: no odor of decomposition.
Class 2: slight odor of decomposition.
Class 3: definite odor of decomposition.

PRODUCT	DEFECT ACTION LEVEL
	Fresh, canned or frozen: Average of 10% rejects due to rot. Average of 4% insect-infested cherries.
Citrus Fruit Juices (canned)	Microscopic mold-count average of 10%. Drosophila and other fly eggs: 5 per 250 mm. Drosophila larvae: 1 per 250 mm. If average of 5% by count have larvae.
Currants Figs	If average of 10% by count are insect-infested and/or show mold and/or dirty fruits or pieces of fruit.
Lingonberries (canned)	Average of 3 larvae per lb.
Multer Berries (canned)	Average of 40 thrips per No. 2 can.
Olives	*Pitted:* Average of 1.3% by count of olives with pit fragments 2 mm. or longer measured in the longest dimension, exclusive of whole pits. *Salad olives:* Average of 1.3 pit fragments per 300 gm., including whole pits and fragments 2 mm. or longer measured in the longest dimension. *Salt-cured olives:* Average of 15% by count of olives with 10 scale insects each; or average of 25% by count of olives showing mold. *Imported black or green:* Average of 10% by count wormy or worm-cut. *Salad-type:* Average of 12% by weight insect-infested and/or in-insect-damaged due to the olive fruit fly.
Peaches (canned)	Average of 5% wormy or moldy fruit by count or 4% if a whole

PRODUCT	DEFECT ACTION LEVEL
	larva or equivalent is found in 20% of the cans.
Pineapple (canned, crushed)	Microscopic mold-count average of 30%.
Plums (canned)	Average of 5% by count of plums with rot spots larger than the area of a circle 12 mm. in diameter.
Prunes (dried)	Average of 10% by count insect-infested and/or showing mold and/or dirty fruits or pieces of fruit.
Prunes (Pitted)	Average of 3% by count prunes with whole pits and/or pit fragments 2 mm. or longer; and 4 or more of the 10 subsamples examined having 3% prunes by count with whole pits and/or pit fragments 2 mm. or longer.
Raisins	Average of 5% by count of natural raisins showing mold. Average of 40 mm. of sand and grit per 100 gm. of natural or Golden Bleached raisins. 10 insects or equivalent and 35 Drosophila eggs per 8 oz. of Golden Bleached raisins.
Strawberries (frozen—whole or sliced)	Microscopic mold-count average of 45% and the mold count 55% in half of the subsamples.

Grains
Popcorn

One rodent pellet in one or more subsamples upon examination of ten 225-gm. subsamples or six 10-oz. consumer-size packages, and one rodent hair in other subs.; or 2 rodent hairs per pound and rodent hairs in 50% of the subs.; or 20 gnawed grains per pound,

PRODUCT	DEFECT ACTION LEVEL
	and rodent hairs in 50% of the subs.
	5% by weight of field corn in the popcorn.
Wheat	One rodent pellet per pint.
	1% by weight of insect-damaged kernels.

Jams, Jellies, Fruit Butters and Fig Paste

Apple Butter	Microscopic mold-count average of 12%.
	Average of 4 rodent hairs per 100 gm.
	Average of 5 whole insects or equivalent (not counting mites aphids, thrips, scales) per 100 gm
Black Cherry Jam	Microscopic mold-count average of 50%.
Black Currant Jam	Microscopic mold-count average of 75%.
Fig Paste	13 insect heads per 100 gm. of fig paste in each of 2 subsamples

Miscellaneous

Corn Husks (for tamales)	If 5% by weight of the corn husks examined are insect-infested (including insect-damaged) or moldy.

Nuts

Tree Nuts	*Nuts in shell and shelled nuts* Reject nuts (insect-infested, rancid, moldy, gummy and shriveled or empty shells) of the following limits, determined by macroscopic examination:

Unshelled	Shelled
Almonds 5%	Almonds 5%
Brazils 10%	Brazils 5%

PRODUCT	DEFECT ACTION LEVEL

Green Chestnuts 15% Cashews 5%
Baked Chestnuts 10% Dried Chestnuts 5%
Filberts 10% Filberts 5%
Pecans 10% Pecans 5%
Pistachios 10% Pistachios 5%
Walnuts 10% Walnuts 5%
Lichee Nuts 15% Pili Nuts 10%
Pili Nuts 15%

Mixed nuts in shell: The above percentage of reject nuts for any one type of nut applies to that type of nut in a mixture. The above limits apply for orchard-type insect infestation.

Peanuts and Peanut Products

Peanuts (shelled or unshelled)

Unshelled: Average of 10% deteriorated or unsound nuts.
Shelled: Average of 5% deteriorated or unsound nuts.
Shelled: An average of 20 insects or equivalent per whole-bag sifting (100-lb-bag basis).

Peanut Butter

Average of 50 insect fragments per 100 gm.
Average of 2 rodent hairs per 100 gm.
Gritty to the taste and the water-insoluble inorganic residue is 35 mg. per 100 gm.

Spices

Allspice

Average of 5% moldy berries by weight.

Bay (Laurel) Leaves

Average of 5% moldy pieces by weight.
Average of 5% insect-infested pieces by weight.

PRODUCT	DEFECT ACTION LEVEL
	Average of 1 mg. of excreta per pound after processing.
Capsicum	*Capsicum Pods:* Average of 3% insect-infested and/or moldy pods by weight.
	Average of 1 mg. of excreta per pound.
	Capsicum Powder: Microscopic mold-count average of 20%.
	Average of 50 insect fragments per 25 gm.
	Average of 6 rodent hairs per 25 gm.
Cassia or Cinnamon (whole)	Average of 5% moldy pieces by weight.
	Average of 5% insect-infested pieces by weight.
	Average of 1 mg. of excreta per pound.
Cloves	Average of 5% stems by weight.
Condimental Seeds, other than Fennel Seeds and Sesame Seeds	Average of 3 mg. of excreta per pound.
Cumin Seed	Average of 9.5% ash and/or 1.5% acid insoluble ash.
Curry	Average of 100 insect fragments per 25 gm.
	Average of 8 rodent hairs per 25 gm.
Fennel Seed	If 20% of subsamples contain excreta and/or insects.
	Average of 3 mg. of excreta per pound.
Ginger (whole)	Average of 3% moldy and/or insect-infested pieces by weight.
	Average of 3 mg. of excreta per pound.

PRODUCT	DEFECT ACTION LEVEL
Hops	Average of 2,500 aphids per 10 gm.
Leafy Spices, other than Bay Leaves	Average of 5% insect-infested and/or moldy pieces by weight. Average of 1 mg. of excreta per pound after processing. Average of 3% insect-infested and/or moldy pieces by weight.
Mace	Average of 3 mg. of excreta per pound. Average of 1.5% foreign matter through a 20-mesh sieve.
Nutmegs	Average of 10% insect-infested and/or pieces showing mold by count.
Pepper (whole)	Average of 1% insect-infested and/or moldy pieces by weight. Average of 1 mg. of excreta per pound. Average of 1% pickings and siftings by weight.
Sesame Seeds	Average of 5% insect-infested or decomposed seeds by weight. Average of 5 mg. of excreta per pound. Average of 0.5% foreign matter by weight.

Vegetables

PRODUCT	DEFECT ACTION LEVEL
Asparagus (canned or frozen)	15% of spears by count infested with 6 attached asparagus beetle eggs or egg sacs.
Beets (canned)	Average of 5% by weight, of pieces with dry rot.
Broccoli (frozen)	Average of 80 aphids or thrips per 100 gm.
Brussel Sprouts (frozen)	Average of 40 aphids and/or thrips per 100 gm.

PRODUCT	DEFECT ACTION LEVEL
Corn (sweet, canned)	Two larvae, cast skins, larval or cast-skin fragments, 3 mm. or longer of corn-ear worm or corn borer, *and* aggregate length of such larvae, cast skins, larval or cast skin fragments exceeding 12 mm. in 24 lbs. (24 No. 303 cans or equivalent).
Greens (canned)	Average of 10% of leaves by count, or weight showing mildew ½ in. in diameter.
Mushrooms (canned)	Average of 20 larvae per 100 gm. of drained mushrooms and proportionate liquid. Average of five larvae, 2 mm. or longer, per 100 gm. of drained mushrooms and proportionate liquid. Average of 75 mites per 100 gm. of drained mushrooms and proportionate liquid. Average of 10% decomposed mushrooms.
Peas, Black-Eyed (cowpeas, field peas—canned)	Average of 5 cowpea curculio larvae or the equivalent per No. 2 can.
Peas, Black-Eyed (cowpeas, field peas—dried)	Average of 10% by count insect damaged.
Peas and Beans (dried)	Average of 5% by count insect infested and/or insect-damaged by storage insects.
Spinach (canned or frozen)	*Canned only:* Average of 60 aphids per 100 gm. of drained spinach and 25% of the sub samples contain 100 aphids per 100 gm. of drained spinach. 2 spinach worms (caterpillars) 3 mm. in length present in 12 No. 2 cans.

PRODUCT	DEFECT ACTION LEVEL
	Canned or frozen: Average of 9 spinach-leaf miners per 100 gm., with half of the spinach-leaf miners 2 mm. in length.
	If average of 10% leaves by count or weight show mildew ½ in. in diameter.

Tomatoes and Tomato Products

PRODUCT	DEFECT ACTION LEVEL
Tomatoes (canned)	10 Drosophila fly eggs per 500 grams; or 5 Drosophila fly eggs and 1 larva per 500 gm.; or 2 larvae per 500 gm.
Tomato Juice	10 Drosophila fly eggs per 100 gm.; or 5 Drosophila fly eggs and 1 larva per 100 gm.; or 2 larvae per 100 gm.
Tomato Puree	20 Drosophila fly eggs per 100 gm.; or 10 Drosophila fly eggs and 1 larva per 100 gm.; or 2 larvae per 100 gm.
Tomato Paste, Pizza and Other Sauces	30 Drosophila fly eggs per 100 gm.; or 15 Drosophila fly eggs and 1 larva per 100 gm.; or 2 larvae per 100 gm.
Tomato Catsup	Microscopic mold-count average of 30%.
Tomato Juice	Microscopic mold-count average of 20%.
Tomato Paste or Puree	Microscopic mold-count average of 40%.
Tomato Sauce (undiluted)	Microscopic mold-count average of 40%.
Canned Tomatoes, with or without added tomato juice	Microscopic mold-count average of 15% of the drained juice.
Canned tomatoes packed in Tomato Puree	Microscopic mold-count average of 25% of the drained packing media.

PRODUCT	DEFECT ACTION LEVEL
Pizza Sauce (based on 6% total tomato solids after pulping)	Microscopic mold-count average of 30%.
Tomato Soup and Other Tomato Products	Microscopic mold-count average of 40%.

NOTES

Introduction

1. Tom Alexander, "The Hysteria about Food Additives," *Fortune*, March 1972.

Chapter 1

1. James S. Turner, *The Chemical Feast: Ralph Nader's Study Group Report on the Food and Drug Administration*. (New York: Grossman paperback, 1970), p. 236. For further discussion of how food standards were weakened by industry lobbying, see pp. 49–66.

2. "Regulations of Food Additives and Medicated Animal Feeds," hearings before a subcommittee of the House Committee on Government Operations, March 1971, p. 276.

3. *Ibid.*, p. 3.

4. Michael F. Jacobson, *Eater's Digest: The Consumer's Factbook of Food Additives* (New York: Doubleday Anchor Books, 1972), p. 12.

5. "Food Additives: What They Are and How They Are Used," Manufacturing Chemists' Association, Washington, D.C.

6. *Food Chemical News*, Oct. 2, 1972.

7. "Nutrition and Human Needs—Food Additives," hearings before the Senate Select Committee on Nutrition and Human Needs, of the United States Senate, Ninety-second Congress, September 1972, p. 1676.

8. *Ibid.*, pp. 1559–60.

9. Tom Alexander, "The Hysteria about Food Additives," *Fortune*, March 1972.

10. Howard J. Sanders, "Food Additives, Part I," *Chemical and Engineering News*, Oct. 10, 1966. For a comprehensive discussion of the food industry's viewpoint toward additives, also see Part II of this series, Oct. 17, 1966.

11. Sidney Margolius, *The Great American Food Hoax* (New York: Walker and Co., 1971), p. 19.

12. *Food Chemical News*, May 1, 1972, p. 39.

13. Testimony of Richard L. Hughes, Senate "Nutrition and Human Needs" hearings, September 1972, pp. 1096–97.

Chapter 2

1. Jean Carper, "Secretary Finch Is Not Alarmed," *The Nation*, March 9, 1970.

2. Daniel Zwerdling, "The Pollution of Food," *Intellectual Digest*, October 1971.

3. *Medical World News*, Oct. 6, 1972.

4. "Chemicals and the Future of Man," hearings before the Subcommittee on Executive Reorganization and Government Research of the Senate Committee on Government Operations, the United States Senate, Ninety-second Congress, April 6 and 7, 1971, p. 173.

5. Beatrice Trum Hunter, *Fact Book on Food Additives and Your Health* (New Canaan, Conn.: Keats, 1972), p. 51.

6. Senate "Nutrition and Human Needs" hearings, September 1972, p. 1560.

7. "Appraisal of the Safety of Chemicals in Foods, Drugs and Cosmetics," published by the Editorial Committee of the Association of Food and Drug Officials of the United States, Topeka, Kans., 1965, pp. 102–4.

8. "Cancer Prevention and the Delaney Clause," report of the Health Research Group, Washington, D.C., 1973, p. 10.

9. Statement of Dr. Samuel Epstein in "Birth Defects and Their Environmental Causes," *Medical World News*, Jan. 22, 1971.

10. Senate "Chemicals and the Future of Man" hearings, p. 172.

Chapter 3

1. *Code of Federal Regulations, Title 21—Food and Drugs*, Parts 1–119 and Parts 120–129 (available from Superintendent of Documents, U. S. Govt. Printing Office, Washington, D.C. 20402; $1.75 per book).

2. James S. Turner, *The Chemical Feast* (New York: Grossman paperback, 1970), pp. 8, 9.

3. House hearings, "Regulation of Food Additives and Medicated Animal Feeds," March 1971, p. 544.

4. Statement of Dr. Samuel Epstein, Senate "Nutrition and Human Needs" hearings, September 1972, p. 1285. The four committee reports are: Crow Report, Genetic Study Section of the National Institutes of Health, 1968; Report of the Advisory Panel on Mutagenicity, Department of Health, Education and Welfare, 1969; Report of the NIEHS Task Force, 1970; Report of the FDA Advisory Committee on Protocols for Safety Evaluation, 1970.

5. For a discussion of the interim regulation, see James S. Turner, "The Delaney Anticancer Clause: A Model Environmental Protection Law," *Vanderbilt Law Review*, Vol. 24, October 1971, p. 897.

6. "Evaluation of Environmental Carcinogens," a report to the Surgeon General, USPHS, by the Ad Hoc Committee on the Evaluation of Low Levels of Environmental Chemical Carcinogens, National Institutes of Health, National Cancer Institute, April 22, 1970.

7. "Nutrition and Human Needs" Senate hearings, September 1972, p. 1288 (statement of Dr. Samuel Epstein).

8. *Ibid.*, pp. 830–34 (statement of Dr. John W. Olney).

9. *Ibid.*, pp. 838–39.

10. *Ibid.*, pp. 1721–31 (statement of Dr. Philip Handler).

Chapter 4

1. Michael F. Jacobson and Robert White, "Company Town at FDA," *Progressive*, April 1973.

2. *Medical World News*, Sept. 22, 1972, p. 6.

3. "Cancer Prevention and the Delaney Clause," Health Research Group report, p. 18. Also see Harrison Wellford, *Sowing the Wind: A Report for Ralph Nader's Center for the Study of Responsive Law on Food Safety and the Chemical Harvest* (New York: Bantam Books, 1973), pp. 184–85.

4. *Food Chemical News*, May 29, 1972.

5. Dr. Charles Edwards, "Benefit and Risk: The Unending Search for Accommodation," Louis Mark Memorial Lecture at the American College of Chest Physicians, Denver, Colo., Oct. 25, 1972.

6. Statement of Dr. Samuel Epstein, Senate "Nutrition and Human Needs" hearings, September 1972, p. 1281.

7. "Evaluation of the Carcinogenic Hazards of Food Additives," fifth report of the Joint FAO/WHO Expert Committee on Food

Additives, World Health Organization Technical Report Series No 220 (1961), p. 19.

8. "Bioassay of Pesticides and Industrial Chemicals for Tumorigenicity in Mice: A Preliminary Note," *Journal of the National Cancer Institute*, Vol. 42, No. 6 (June 1969), pp. 1101–14.

9. Peter Barton Hutt, "Safety Regulation in the Real World," remarks at the National Academy of Sciences' Forum on the Design of Policy on Drugs and Food Additives, May 15, 1973.

10. "Lindsay Stresses FDA Concern with Benefit–Risk Ratios," *Food Chemical News*, Feb. 1, 1971.

11. Statement of Dr. Umberto Saffiotti, Senate "Chemicals and the Future of Man" hearings, April 1971, pp. 176–77.

Chapter 5

1. The two studies are: A. I. Shtenberg and E. V. Gavrilenko "Effect of Amaranth Food Dye on Reproductive Function and Progeny Development in Experiments with Albino Rats, *Vop. Pitan*, 29, 66 (1970), and M. M. Andrianova, "Carcinogenic Properties of the Red Food Dyes Amaranth, Ponceaux Sx, and Ponceaux 4R," *Vop Pitan*, 29, 61 (1970).

2. Jean Carper, "Food Coloring," *News and Opinion*, May 24 1972.

3. "Red Food Coloring: How Safe Is It?," *Consumer Reports* February 1973.

4. *Ibid*.

5. Statement of Anita Johnson, "Nutrition and Human Needs" hearings, September 1972, p. 1563.

6. *Medical World News*, Sept. 8, 1972.

Chapter 6

1. Personal communication.

2. "Regulation of Food Additives—Nitrites and Nitrates," 19th report of the House Committee on Government Operations, Aug. 15, 1972, p. 11.

3. A. J. Lehman, "Nitrates and Nitrites in Meat Products," *Journal of the Association of Food and Drug Officials of the United States*, Vol. 12 (1958), pp. 136–38.

4. Statement of Dr. William Lijinsky, House hearings "Regulation of Food Additives and Medicated Animal Feeds," March 1971, p. 132.

5. N. P. Sen, D. C. Smith, and L. Schwinghamer, "Formation of

N-nitrosamines from secondary amines and nitrite in human and animal gastric juice," Fd. Cosmet. Toxicol. 7:301–07, 1969.

6. For complete list, see House hearings, "Regulation of Food Additives and Medicated Animal Feeds," March 1971, pp. 15–128.

7. "How Sodium Nitrite Can Affect Your Health," report by Michael F. Jacobson, Ph.D., Center for Science in the Public Interest, March 1973.

8. "The FDA and Nitrite," by Dale Hattis, Stanford Univ. School of Medicine, April 1972, mimeographed report distributed by the Environmental Defense Fund, Washington, D.C.

9. For a discussion of this subject, see House hearings "Regulation of Food Additives and Medicated Animal Feeds," March 1971, pp. 556–63.

10. *Ibid.*, p. 210 (testimony of Dr. Charles Edwards).

11. "Regulation of Food Additives—Nitrites and Nitrates," House Government Operations Committee report, p. 22.

Chapter 7

1. Court decision in *Bell v. Goddard*, U. S. Court of Appeals, 7th Circuit, Aug. 11, 1966, as cited in "Regulation of Diethylstilbestrol (DES)," Part II, hearings before a subcommittee of the House Committee on Government Operations, Dec. 13, 1971, pp. 195–200.

2. Quoted in Jean Carper, "Danger of Cancer in Food," *Saturday Review*, Sept. 5, 1970.

3. "Regulation of Diethylstilbestrol (DES)," hearings before the Subcommittee on Health of the Senate Committee on Labor and Public Welfare, United States Senate, Ninety-second Congress, Second Session on S. 2818, July 20, 1972, p. 9 (statement of Senator William Proxmire).

4. House hearings "Regulation of Diethylstilbestrol (DES)," Part I, Nov. 11, 1971, p. 36 (testimony of Dr. Umberto Saffiotti).

5. *Ibid.*, p. 67 (testimony of Dr. Roy Hertz).

6. See *ibid.*, Part II, Dec. 13, 1971, pp. 113–46.

7. House hearings "Regulation of Food Additives and Medicated Animal Feeds," March 1971, pp. 243–44.

Chapter 8

1. Transcript of proceedings of the Interagency Meeting on Polychlorinated Biphenyls (PCBs), Dept. of Health, Education and Wel-

fare, Aug. 5, 1971. This document is an excellent source on the toxicity of PCBs.

2. Quoted in Hank Burchard, "Millions of Chickens Tainted," *Washington Post*, July 29, 1971.

3. "The Toxic Substances Control Act of 1971 and Amendment," hearings on S. 1478 before the Subcommittee on the Environment of the Senate Committee on Commerce, August–November 1971, p. 110 (statement of Harrison Wellford).

4. *Ibid.*, p. 99 (letter from Sam Spencer to the subcommittee).

5. *Ibid.*, p. 100 (testimony of Robert Risebrough); for additional discussion of the PCB hazard by Risebrough, see p. 96.

6. Hank Burchard, "Chickens Fed Tainted Meal Termed Safe," *Washington Post*, July 30, 1971.

7. "The High-Filth Diet, Compliments of FDA," *Consumer Reports*, March 1973.

8. "Plant Pathogens: A New Harvest of Suspicion," *Medical World News*, Sept. 22, 1972.

Chapter 9

1. Interview on station KDKA-TV, Pittsburgh, Sept. 4, 1969.

2. For the complete text of these memos, see "Cyclamate Artificial Sweeteners," hearings before a subcommittee of the House Committee on Government Operations, June 10, 1970, pp. 15–23.

3. *Ibid.*, p. 51 (testimony of William Goodrich).

4. Jean Carper, "Secretary Finch Is Not Alarmed," *The Nation*, March 9, 1970.

5. Quoted in "Cyclamate Artificial Sweeteners" hearings, pp. 68-69.

6. *Washington Post*, Jan. 29, 1970.

BIBLIOGRAPHY AND SUGGESTED READINGS

Much of the information in this book came from the personal knowledge of Dr. Jacqueline Verrett and from personal investigations by Jean Carper. However, there are a number of Congressional-hearing records which were important sources, and other publications which shed light on the hazards of chemicals in food and the FDA's participation, some of which are not mentioned in the Notes. We would like to acknowledge the following as helpful in our research and as recommended reading for consumers.

Books

Hunter, Beatrice Trum, *Consumer Beware*. New York: Simon and Schuster, 1971.

———, *Fact Book on Additives and Your Health*. New Canaan, Conn.: Keats Publishing Co., 1972.

Jacobson, Michael F., *Eater's Digest: The Consumer's Factbook on Food Additives*. New York: Doubleday Anchor Books, 1972.

Marine, Gene, and Judith Van Allen, *Food Pollution*. New York: Holt, Rinehart and Winston, 1972.

Turner, James S., *The Chemical Feast: Ralph Nader's Study Group Report on the Food and Drug Administration*. New York: Grossman (hardcover and paperback), 1970.

Wellford, Harrison, *Sowing the Wind: A Report for Ralph Nader's Center for Study of Responsive Law on Food Safety and the Chemical Harvest*. New York: Grossman (hardcover), 1972; Bantam (paperback), 1973.

Congressional Hearings

"Chemicals and the Future of Man," hearings before the Subcommittee on Executive Reorganization and Government Research of the Committee on Government Operations, U. S. Senate, 92nd Congress, April 6 and 7, 1971 (chaired by Sen. Abraham A. Ribicoff).

"Cyclamate Artificial Sweeteners," hearings before a subcommittee of the Committee on Government Operations, House of Representatives, 91st Congress, 2nd session, June 10, 1970 (chaired by Rep. L. H. Fountain).

"Nutrition and Human Needs—Food Additives," Parts 4A, 4B and 4C, hearings before the Select Committee on Nutrition and Human Needs, U. S. Senate, 92nd Congress, 19, 20, 21, 1972 (chaired by Sen. Gaylord Nelson).

"Regulation of Diethylstilbestrol (DES)," hearings on S. 2818 before the Subcommittee on Health of the Committee on Labor and Public Welfare, U. S. Senate, 92nd Congress, 2nd session, July 20, 1972 (chaired by Sen. Edward M. Kennedy).

"Regulation of Diethylstilbestrol (DES)," Parts I and II, hearings before a subcommittee of the Committee on Government Operations, House of Representatives, 92nd Congress, Nov. 11 and Dec. 13, 1971 (Fountain).

"Regulation of Food Additives and Medicated Animal Feeds," hearings before a subcommittee of the Committee on Government Operations, House of Representatives, 92nd Congress, March 16, 17, 18, 29 and 30, 1971 (Fountain).

"The Toxic Substances Control Act of 1971 and Amendment," hearings on S. 1478 before the Subcommittee on the Environment of the Committee on Commerce, U. S. Senate, 92nd Congress, 1st session, Aug. 3, 4, 5, Oct. 4 and Nov. 5, 1971.

Reports

"Cancer Prevention and the Delaney Clause," a report by the Health Research Group, 2000 P. Street, N.W., Washington, D.C., January 1973, funded by Ralph Nader's Public Citizen.

"The FDA and Nitrite," a case study of violations of the Food, Drug and Cosmetic Act with respect to a particular food additive, mimeographed report by Dale Hattis, Dept. of Genetics, Stanford University School of Medicine, April 25, 1972, sponsored and distributed by the Environmental Defense Fund, Washington, D.C.

"How Sodium Nitrite Can Affect Your Health," report by Michael
    F. Jacobson, Ph.D., Center for Science in the Public Interest,
    1779 Church Street, N.W., Washington, D.C. 20036, February
    1973 ($2).
"Regulation of Food Additives—Nitrites and Nitrates," 19th report of
    the Committee on Government Operations, House of Rep-
    resentatives, Aug. 15, 1972, based on a study by the Intergovern-
    mental Relations Subcommittee (a Fountain committee report).

Newsletters

Food Chemical News, ed. by Lou Rothschild, 1341 G Street, N.W.,
    Washington, D.C. 20005.
Newsletter of the Federation of Homemakers, Inc., ed. by Mrs.
    Ruth Desmond, 927 North Stuart Street, Arlington, Va. 22203.

INDEX